The Potential Effects of Ozone Depletion in the United Kingdom

LONDON: The Stationery Office

ISBN 0117533130

Cover photograph:
This view of the Sun shows the solar chromosphere which lies between the sunspot layer (Photosphere) and the Corona. In this view prominences appear as dark filaments seen against the disc and the bright active regions show up very clearly.
(Photograph: © NASA/JPL, from The Planetarium, Armagh, N. Ireland)

Produced by the Department of the Environment
Printed in Great Britain on recycled paper
Designed by DDP Services
B8758
October 1996

United Kingdom UVB Measurements and Impacts Review Group

A Webb *(Chairperson)*	University of Manchester Institute for Science and Technology
B Gardiner	British Antarctic Survey (BAS)
C Driscoll	National Radiological Protection Board (NRPB)
J Austin	Meteorological Office
C Arlett	University of Sussex
J Raven	University of Dundee
D Lowe	Plymouth Marine Laboratory
N Paul	University of Lancaster
A McCleod	Institute of Terrestrial Ecology
B Diffy	Dryburn Hospital
A Young	St Thomas's Hospital
J E Hillerton	Institute of Animal Health
D Gardiner	Building Research Establishment
G Bentham	University of East Anglia
P Bramwell *(Executive Secretary)*	Department of the Environment

Contents

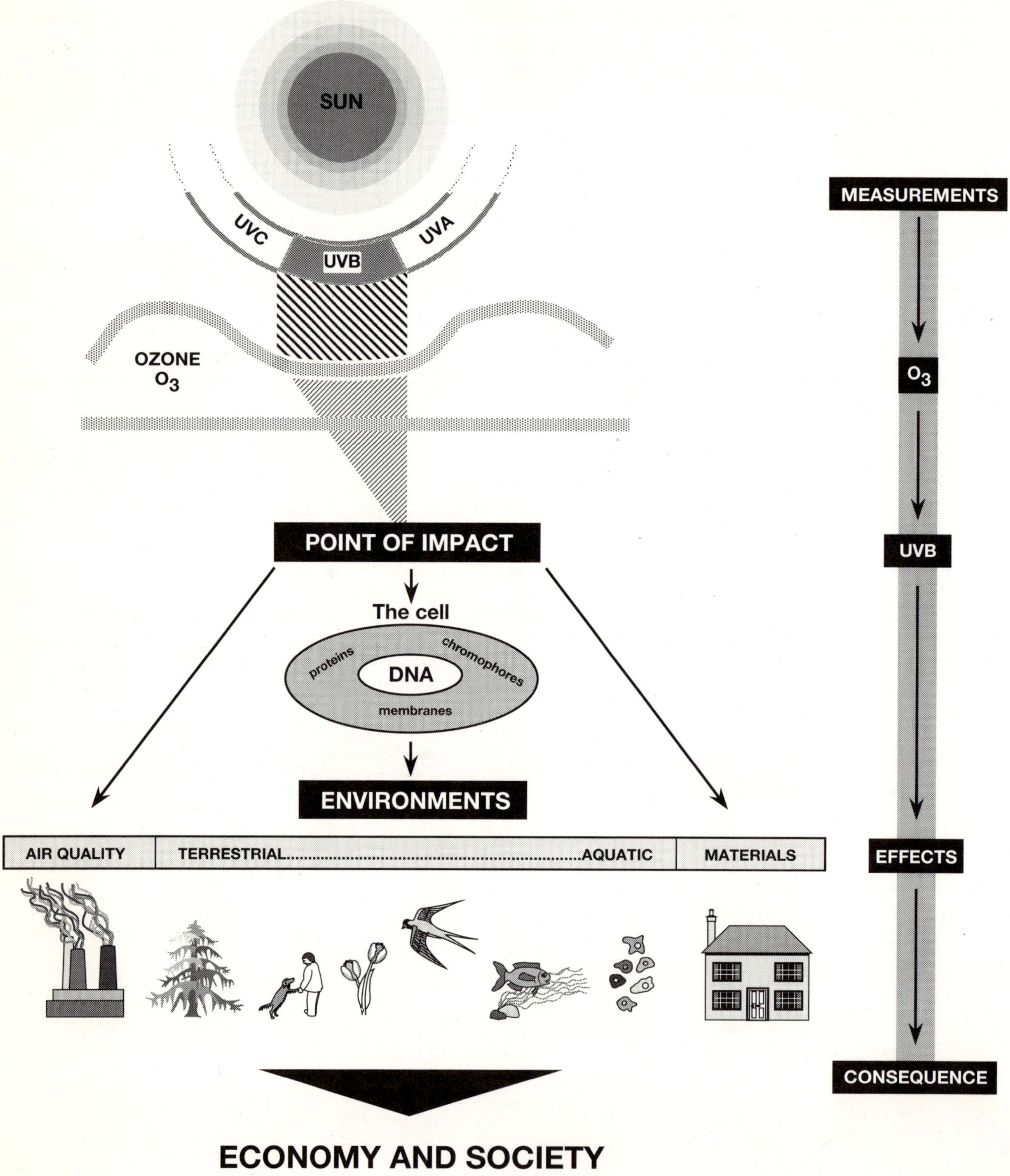
SUN
UVC
UVB
UVA
OZONE
O_3
POINT OF IMPACT
The cell
proteins
chromophores
DNA
membranes
ENVIRONMENTS
AIR QUALITY
TERRESTRIAL......AQUATIC
MATERIALS
ECONOMY AND SOCIETY
MEASUREMENTS
O_3
UVB
EFFECTS
CONSEQUENCE

Preface

Environmental matters are particularly challenging to scientists and policymakers alike. For scientists they often involve complex interactions and links which require multidisciplinary approaches. The issue of stratospheric ozone depletion as a result of man made halocarbons requires an understanding of the processes of meteorology, atmospheric physics, chemistry and various aspects of biology and medicine, in a global context.

Increases in ultraviolet radiation resulting from stratospheric ozone depletion enhance the risk of damage across a wide range of sectors ranging from human health, flora and fauna, agriculture, materials and air quality. The socioeconomic consequences of these also have to be considered.

To date research on the impacts of elevated UVB as a result of ozone depletion has been carried out in a fairly fragmented way, with little contact between researchers from different disciplines.

In order to assess the overall risk of ozone depletion in the UK, the Department of the Environment convened the Ultraviolet Measurements and Impacts Review Group (UMIRG) in 1995 to draw on the wide range of work being undertaken and provide the best scientific assessment of UVB measurements and impacts in the United Kingdom, both now and in the future. The terms of reference for UMIRG are reproduced at annex 1.

The Department has convened another related Review Group, the Stratospheric Ozone Review Group (SORG) which was set up by Department of the Environment and the Meteorological Office in 1985 to provide advice and inform the policy process on the scientific issues associated with the ozone layer and its depletion. The SORG is reporting at the same time as the UMIRG.

In its 1996 report, SORG notes that over northern mid-latitudes, an annually averaged decrease in total ozone of 4 to 5% per decade has taken place since 1979. The greatest losses are observed during spring and record low amounts of ozone were observed over the UK in March 1996. The SORG also notes that large ozone losses in high northern latitudes are likely to continue for at least the next decade in cold winters.

It is therefore timely to assess the risks associated with ozone depletion in the UK over the next few years and also to consider the implications of an unanticipated further loss of ozone.

Executive Summary

One of the principal reasons for the widespread concern about depletion of the stratospheric ozone layer is the anticipated increase in the levels of ultraviolet (UV) radiation received at the surface of the Earth, and the consequential effects on our environment and well-being. Information on changes in UV radiation and its effects in the UK are limited. This assessment has been prepared to review what is currently known so that appropriate actions can be planned by Government and others, should present measures to control damage to the ozone layer prove less than effective.

- The ozone layer shields the Earth from much of the Sun's harmful ultraviolet radiation. Global levels of ozone have declined in recent years. Since the late 1970s column ozone[1] losses of about 10% have occurred over the UK and northern Europe during the winter and spring, whereas losses in summer and autumn have been about 5%.
- Concentrations of ozone depleting substances in the lower atmosphere are beginning to stabilise in response to the Montreal Protocol but are not expected to decline to pre-ozone hole levels until the middle of the next century. Although there are still uncertainties in our understanding of ozone loss processes, especially over mid-latitudes, stratospheric ozone amounts are likely to begin to recover following the anticipated fall in chlorine and bromine levels in the atmosphere.
- There is no doubt that occurrences of high UV are strongly correlated with low ozone. UV reaching the ground depends upon several atmospheric factors besides ozone e.g. aerosol and clouds. To date, long term trends in UV have not been detected, as high quality data records do not cover a sufficient period to distinguish small trends from the natural variability. Nevertheless, the accumulated UV dose must have increased beyond that which would have been received with no ozone depletion.
- Increased exposure to UV radiation will increase the likelihood of deleterious effects on biological systems. In humans these effects include sunburn and skin cancers, together with possible eye damage and changes in immune function. Risk is related to exposure and can therefore be reduced by behavioural responses. In land and aquatic animals similar effects can be expected. Management strategies could mitigate the risks in domestic and farmed animals, but wildlife may be vulnerable if unable to adapt to the timescale and magnitude of UV changes.
- Increased UV radiation can modify the development of flora and fauna, and interactions between species, with implications for biodiversity in UK aquatic and terrestrial ecosystems. Whilst evidence is sparse, primary production may be compromised in both aquatic and terrestrial food chains. Possible reductions in agricultural yield could be accommodated by plant

[1] Column ozone: The total amount of ozone in a vertical column through the atmosphere.

breeding for increased UV tolerance in annual crops, but this has limited practical value for perennial species and natural vegetation.

- Natural and synthetic polymers, widely used by the construction, transport and agricultural industries are degraded by exposure to sunlight. UV radiation can be particularly damaging. At first the appearance may be affected but in the longer term the physical integrity may be compromised so that the material no longer performs as intended. Nevertheless, the UV stability of most common polymers can be improved with currently available techniques.
- Increased UV as a result of stratospheric ozone depletion affects the chemistry of the lower atmosphere. This is expected to reduce air quality by increasing surface ozone concentrations in urban areas and downwind, which could counteract in part the increased UV that would otherwise reach the surface.
- International action has been taken to reduce ozone depletion but the damage caused to the ozone layer by man-made chemicals will not be resolved for many decades. Socio-economic implications of enhanced exposure to UV radiation include the costs of treatment and prevention programmes associated with any health effects. There will also be financial burdens associated with impacts on agriculture, forests, fishing and materials.

It is clear that UV radiation can have significant effects on many processes. However at present serious gaps in our knowledge make it impossible to quantify precisely the damage resulting from decreased ozone levels. If the commitments under the Montreal Protocol are fulfilled and other atmospheric factors do not exacerbate chemical ozone depletion, the increase in UV over the next few decades should be temporary and the resultant damage containable.

Future Research

Our lack of knowledge in many solar UV related areas needs to be addressed, not only against the possibility of large changes in UV, but also to assess the impacts of the small but long-term changes which are already occurring.

1. Long-term monitoring and prediction of changes in ozone and UV is a prime requirement. Current observational programmes, and physical and biological dosimetry, are essential to provide the data from which to determine a more detailed knowledge of UV exposure patterns.
2. Action spectra[2] for critical end-points in human and animal health, plant growth, and long-term ecosystem processes must be defined to enable a quantitative assessment of possible impacts to be carried out.
3. General conclusions concerning future ecosystem responses, and particularly the development of predictive models, require knowledge of the biological dose responses of UV made using the best available methodologies.
4. Further information is needed on the action of UV on man and animals particularly identification of skin cancer risk factors, effects on immune function, and physiological and behavioural adaptive responses.
5. The extent and mechanisms of UV effects on ecological processes, primary production, biodiversity and interactions of UV with other environmental changes are required for assessment of impacts on aquatic and terrestrial ecosystems.
6. Procedures need to be established for measuring the sensitivity of practical, polymer-based materials to increased UV levels so that vulnerable materials can be identified.

---

[2] Action Spectra: Refer to Box 1 in chapter 1.

1. Introduction

- **The ozone layer is essential to protect life on Earth from the harmful effects of the Sun's ultraviolet radiation. Global levels of ozone have declined in recent years, and severe ozone depletion is now an annual event in the Antarctic.**
- **International action has been taken to solve these problems, but the damage to ozone caused by man-made chemicals will not be resolved for many decades. The potential effects of the increased ultraviolet radiation which results from ozone depletion should be examined, in case the remedial measures fail.**
- **The powerful influence of ozone depletion on ultraviolet radiation fluxes has already been observed in the Antarctic, but ozone is not the only factor determining UV radiative transfer through the atmosphere. Clouds and other atmospheric constituents also affect the amount of ultraviolet radiation reaching the ground.**
- **The damaging effect of ultraviolet radiation depends on its intensity and its wavelength. Each biological process has its own spectral response, which determines its susceptibility to ozone depletion.**

Ultraviolet radiation

The light and heat that we receive from the Sun are forms of radiation which belong to a large class of emissions known as the electromagnetic spectrum. Radiowaves, microwaves, infra-red, visible light, ultraviolet, X-rays and gamma rays all belong to this category, with each type covering its own range of wavelengths in the vast spectrum of electromagnetic radiation.

Ultraviolet radiation (UV) is the region of the electromagnetic spectrum adjacent to that of visible light, lying beyond the shortest visible wavelengths which are to be found at the violet end of the visible spectrum. The boundary between UV and visible radiation is conventionally set at a wavelength of 400 nm. (A nanometre (nm) is a thousand-millionth of a metre.) UV radiation can be subdivided into three wavebands called UVA (the wavelength range 315 to 400 nm), UVB (280 to 315 nm) and UVC (200 to 280 nm). The Sun emits UV radiation across the full spectral range, but the solar UV spectrum at the Earth's surface does not extend below about 290 nm owing to absorption by oxygen in the upper atmosphere, stratospheric ozone, and scattering and absorption processes in the troposphere.

Although the total energy contained in the UVB part of the solar spectrum is very small (less than 1%), the individual photons are the most energetic to reach the troposphere and are capable of initiating many biological and chemical reactions. It is the results of these reactions, many of which are detrimental, which have caused much of the concern about declining levels of ozone and the resulting increases in UV radiation that might be expected.

Without the protective effect of atmospheric ozone, the surface of the Earth would be exposed to doses of ultraviolet radiation from the Sun which would be lethal for almost all forms of life, including bacteria, plants, animals and humans.

Ozone depletion

In recent years the ozone layer has suffered severe damage in Antarctica, and a gradual decline in ozone has also been detected in other parts of the world. International action has been taken to halt and reverse these trends, but it will be some years before the effect is seen, and many decades before the damage is repaired. In the meantime it is prudent to consider the consequences of the increase in ultraviolet radiation which would result if the remedial measures were to prove ineffective. This could arise either through serious breaches or circumvention of the international agreements, or simply through deficiencies in our understanding of the complex chemistry of the atmosphere.

Fortunately, the most dramatic depletion of ozone has so far been confined to the Antarctic. Every year, as daylight returns after the winter darkness, the total amount of ozone in the atmosphere over Antarctica falls to only 40% of its normal pre-1970 value. This process of ozone destruction, which takes about six weeks during the Antarctic spring, leads to the formation of the so-called ozone hole. At altitudes between 13 and 22 km, in the heart of the ozone layer, almost all the ozone is removed (SORG, 1993). It is now accepted that Antarctic ozone is destroyed by catalytic cycles involving chlorine and bromine released from man-made halocarbons. Antarctic ozone depletion has become progressively worse since its inception in the 1970s, as the chlorine loading of the stratosphere has increased. The Montreal Protocol on Substances that Deplete the Ozone Layer seeks to solve the problem by controlling the production and consumption of halocarbons.

But ozone depletion has also been observed outside the Antarctic during the same period. Between 1979 and 1991 there was a downward trend of about 4% per decade in the average amount of ozone over mid-latitudes in both hemispheres. If the period up to 1994 is considered, the trend in northern mid-latitudes is about 1% per decade more severe, but this may be due to the temporary additional effect of debris from the 1991 eruption of Mt. Pinatubo.

The rate of decrease of ozone is not constant throughout the year in northern mid-latitudes, being rather steeper in the winter and spring than in the summer and autumn.

In the winters of 1991-92 and 1992-93, unusually low ozone values were recorded over northern Europe. These are largely attributable to the natural effects of stationary anticyclones in redistributing ozone, and are not therefore indicative of an increased long-term trend. There is now strong observational evidence for significant chemical destruction of ozone during recent winters in the northern high latitudes, but ozone values outside the Arctic vortex in the 1994-95 winter were consistent with the observed long-term decrease.

Future ozone trends over Europe are therefore difficult to estimate, but they are not likely to improve significantly until the effects of the Montreal Protocol are seen.

Atmospheric absorption of UV radiation

Ozone is a strong absorber of ultraviolet radiation, especially at the shorter wavelengths, hence its valuable protective effect. It is already well established that ozone depletion gives rise to higher levels of ultraviolet radiation at the surface of the Earth, other things being equal. Measurements at the South Pole show a clear anti-correlation between UVB irradiance and ozone measured by satellite, during a period with little variation due to cloud cover. Noon values of biologically-weighted UV radiation recorded at Palmer Station, Antarctica during the annual ozone hole can exceed the summer maximum measured at San Diego, California, despite the much lower solar elevations encountered in Antarctica. (SORG, 1993; WMO, 1995)

However, ozone is by no means the only factor determining the flux of ultraviolet radiation reaching the ground. The elevation of the Sun above the horizon, the cloud regime, the aerosol content of the atmosphere, and the surface reflectivity, all play important parts.

Ultraviolet photons are scattered by air molecules, some returning to space as a result. Tropospheric pollution can scatter and absorb UV radiation, while clouds play a complex role by reflecting some radiation to space and altering the path of the transmitted radiation as it passes through the ozone in the troposphere. Other gases such as nitrogen dioxide and sulphur dioxide also absorb UV, although their effect is much less than that of stratospheric ozone, simply because their concentrations are much smaller. Finally, radiation which is reflected by the Earth's surface may be scattered back downwards again by clouds or other constituents of the atmosphere, so that the UV irradiance may be enhanced by a high surface reflectivity.

It follows that future trends in UV radiation depend to some extent on any systematic changes that may arise in the effective cloud cover or the other factors mentioned above.

Wavelength

It is important to realise that the intensity of solar radiation and its transmission through the atmosphere are both very dependent on wavelength. At the Earth's surface, the solar spectral irradiance at 400 nm is typically about a thousand times greater than in the region near 300 nm. Air molecules scatter radiation in inverse proportion to the fourth power of the wavelength, and ozone absorption has an even steeper spectral variation.

Furthermore, most biological effects of UV radiation are also strongly dependent on wavelength. In general, the shorter the wavelength, the greater will be the biological effect, in accordance with the physical principle that the energy carried by a photon is inversely proportional to its wavelength. However, the relative effectiveness of the different wavelengths is not the same for all biological processes. In fact, each process has its own relative spectral effectiveness, which can be expressed as a function of the wavelength. This function is known as the action spectrum for that biological (or chemical) process (Box 1), and is usually normalised to unity at an arbitrary wavelength such as 300 nm.

As the net result of incoming radiation (the Effective Radiation, see Box 1) is the convolution of the spectral irradiance and the appropriate action spectrum, the effect of a change in ozone will be different for each process. For example, if the action spectrum stops at 310 nm, longer wavelengths having no effect, then the process will be particularly sensitive to changes in ozone, which absorbs strongly at short wavelengths. By contrast, if the action spectrum extends undiminished to 400 nm, then the process will hardly be affected by ozone variations, as the biological effect will be mainly due to the stronger flux at the longer wavelengths where ozone does not absorb.

Most of the action spectra discussed in the context of solar UV radiation refer to biological effects. Of these, the most commonly used is the action spectrum for erythema (sunburn). Many broadband instruments are designed to incorporate this action spectrum weighting, so that they aim to measure erythemally effective radiation directly.

As a rough guide to the importance of ozone variations for each type of biological effect, it is possible to calculate a so-called Radiation Amplification Factor (RAF), which represents the ratio of the fractional increase in the effective radiation to the fractional decrease in ozone (Box 2). Strictly speaking, these factors will themselves be functions of the solar elevation, the ozone amount, and the other atmospheric parameters which have a spectral effect, but if they are calculated for typical values of ozone and solar elevation, then they provide an adequate indication of the sensitivity of the process to ozone changes, bearing in mind that the action spectra are generally poorly known. Some representative RAFs are shown in Table 1.1 for a range of biological and chemical processes.

Table 1.1

Radiation Amplification Factors

Process	RAF	Reference
DNA:		
Generalized DNA damage	2.2	Setlow, 1974
Plants:		
Generalized plant spectrum	2.0	Caldwell *et al.*, 1986
Photosynthetic electron transport	0.2	Bornman *et al.*, 1984
Skin:		
Erythema	1.1	McKinlay & Diffey, 1987
Skin cancer	1.2	de Gruijl & van der Leun, 1994
Eyes:		
Damage to cornea	1.2	Pitts *et al.*, 1977
Damage to lens (cataract)	0.8	Pitts *et al.*, 1977
Materials:		
Yellowing in PVC	0.2	Andrady *et al.*, 1989
Yellowing in polycarbonate	0.4	Andrady, 1991

Source: Madronich et al., 1995

The RAF represents the change in effective radiation due to a change in ozone, but the biological effect may not be proportional to the effective radiation. In this case, a Biological Amplification Factor (BAF) can be defined, to represent the ratio of the fractional increase in the biological effect to the fractional increase in the effective radiation (Box 2). BAFs are very difficult to determine and are usually based on epidemiological (skin cancer) or laboratory results which cannot necessarily be extrapolated for wider use.

The following chapters describe the current state of the art and our understanding of UV measurements and effects, with particular emphasis on research relevant to the UK.

Figure 1.1

Biological action spectra for three UV effects

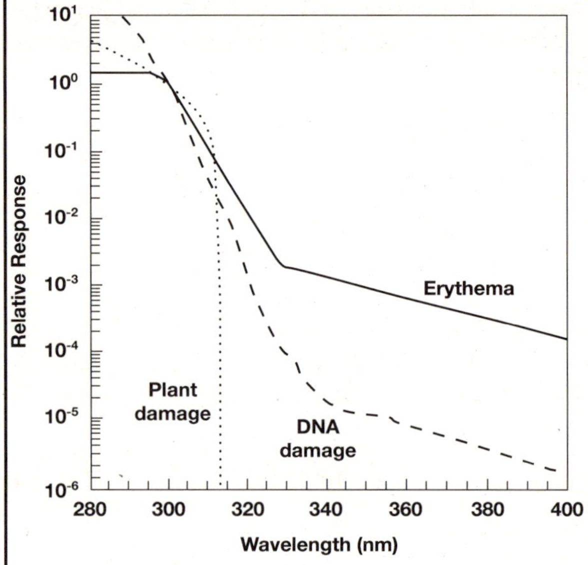

Box 1

Action Spectrum

Photo-Biological and chemical responses to radiation are very dependent on the incident wavelengths of that radiation. For most processes, the shorter wavelengths are more effective, while the longest UV wavelengths may have little or no effect. For a particular biological effect, the relative importance of each wavelength can be expressed in the form of a function known as the action spectrum. For example, if the action spectrum had a value of 1.0 at 300 nm, and 0.5 at 320 nm, then twice as much energy would be required at 320 nm as at 300 nm to have the same effect. Most action spectra are defined to have a value of 1.0 at some arbitrary wavelength, usually either at 300 nm or at the wavelength of maximum effect. As action spectra are notoriously difficult to determine experimentally, they are often adopted more for the sake of international uniformity than for their inherent accuracy. Furthermore, the true action spectrum for a particular biological effect will generally vary considerably from case to case even within a single species, owing to variations in the organisms themselves and in the environmental conditions. Some examples of widely-used action spectra are shown in Figure 1.1.

Effective Radiation

The effective radiation for a particular biological (or chemical) effect is given (in units of "effective W m^{-2}", for example) by

$$F = \int I(\lambda)\ A(\lambda)\ d\lambda$$

where $I(\lambda)$ is the incident spectral radiation

$A(\lambda)$ is the action spectrum for the effect

and λ is the wavelength

The effective radiation dose is given (in units of "effective J m^{-2}", for example) by

$$D = \int F\ dt$$

Box 2

Radiation Amplification Factor

The RAF is the ratio of the fractional increase in the effective radiation to the fractional decrease in ozone. Thus

$$dF/F = -r\, dA/A$$

where F is the effective radiation

A is the amount of ozone

and r is the RAF

from which it follows that a value for the RAF can be obtained from the observed ozone and effective radiation on two separate occasions, from

$$r = - \ln (F_1/F_2) / \ln (A_1/A_2)$$

This assumes that the RAF is invariant over the range of conditions encountered. In practice, the variations of the RAF with ozone amount, solar elevation and other atmospheric conditions are small when compared with the inherent uncertainty arising from imperfect knowledge of the action spectra for the various biological effects. It should be recognised that the concept of an RAF is not very precise, but it nevertheless provides a worthwhile indication of the sensitivity of a biological process to a change in the ozone amount.

Biological Amplification Factor

The BAF is the ratio of the fractional increase in the biological effect to the fractional increase in the effective radiation. Thus

$$dE/E = b\, dF/F$$

where E is the observed biological effect

F is the effective radiation

and b is the BAF

Both the way in which F is measured and the way in which E is quantified can change the calculation of a BAF. This is a particular problem with biological systems where an effect may manifest itself only after many years of exposure (e.g. skin cancer), or where there are many ways to assess damage (e.g. dry weight, yield, height, for plant growth). Often BAFs are based on laboratory studies with animal models or controlled growth chambers, and are then used to extrapolate to wider populations and the real environment.

2. Measurements and Predictions of Ozone and UV

- **Ozone depletion has occurred globally outside the tropics. During winter and spring, column ozone has been depleted by about 10% over the UK and northern Europe since the late 1970s, with smaller losses during summer and autumn. Further losses may be expected over the next decade, but precise predictions are uncertain because of a poor understanding of all the atmospheric coupling processes involved.**
- **There is a continuing need for accurate and continuous measurements of column ozone and solar UV radiation at the earth's surface to establish reliable data records and to identify any changes which may occur.**
- **Short-term changes in UV have been observed at a number of locations and correlated with low ozone events. However, determining a trend in UV is more difficult than determining a trend in ozone and requires long-term measurements. To date UV measurements in the UK have indicated no significant trend in annually integrated UV. Nonetheless, current UV doses are expected to be higher than they would have been without ozone depletion.**
- **UV forecasts for clear skies have been produced for the UK public since the summer of 1994, to provide warnings of sunburn risk.**

2.1. Introduction

In this section we are concerned with the UV radiation at the ground and the 'column ozone'. Column ozone is a measure of the total number of ozone molecules above the ground and is influenced on short spatial and temporal scales by movements in atmospheric pressure systems and on longer timescales by photochemical processes. In the lower atmosphere below altitudes of about 10 km, i.e the troposphere, poor air quality can result in an increase in local ozone and hence in the total vertically integrated column (see Chapter 9). Some 90% of the total column, however, resides in the stratosphere at altitudes above 10 km. At these altitudes the ozone column is also influenced by Stratospheric winds and photochemistry. In high latitudes column ozone is generally high during the spring because stratospheric winds bring ozone from the high production region above the tropics. Column ozone has been measured (in Dobson Units, DU) from the ground since the 1930s, initially at only a few sites but since 1957 on a more widespread and organised basis. Truly global measurements of column ozone first became available with the advent of satellite instrumentation in the 1970s. Since then ozone has been measured from above and below by a number of different instruments.

The record of UV measurements is neither as long, organised or comprehensive as those for ozone. Routine measurements did not begin until there was a perceived need to know about possible changes in UV. Over the last 20 years concerns relating to possible depletion of the stratospheric ozone layer and UV induced skin cancer led to several national and a few international terrestrial UV radiation measurement programmes being established, including measurements in North America, Europe and Australasia, as well as in polar regions. However, it was not until definitive proof of ozone depletion was available that UV measurement programmes really began to expand. Generally, the measurement programmes set up prior to 1990 have not been coordinated internationally and provide only a limited database on personal exposure to solar UV for epidemiological studies. In addition, few programmes have involved continuous measurements over long periods; most measurements being made sporadically on a range of instruments with little or no interlaboratory comparisons. This situation has begun to change and the last five years have seen the start of several monitoring programmes aimed at long term measurements, and a number of international intercomparison events.

2.2. Ozone depleting chemicals

The concentration of ozone in the atmosphere depends on the balance between production and destruction by natural and man-made constituents. Natural constituents which destroy ozone include products from water vapour and nitrous oxide. The latter has been enhanced by the increased use of fertilizers, while some ozone destruction results from naturally occuring chlorine. Ozone depletion by man-made constituents, however, has focused on two other classes of chemical. Chlorofluorocarbons (CFCs) and brominated hydrocarbons were first extensively used in the 1970s and although emitted locally into the atmosphere, over a period of several years they become well mixed globally throughout the lower atmosphere. The CFCs are initially inert but in time are carried by atmospheric motions to high altitudes above the ozone layer. Here the molecules are stripped of the halogen atoms by short wavelength UV radiation. These atoms are then brought down to lower levels in high latitudes where ozone destruction can commence by catalytic processes involving chlorine and bromine.

Predictions of the future ozone amounts are indicated in section 2.9 assuming that future chlorine and bromine concentrations and greenhouse gases evolve in accordance with the Montreal Protocol and amendments. These predictions are, however, uncertain for several reasons. Failure to comply globally with the Montreal Protocol agreements may lead to severe ozone depletion which cannot be quantified at present. Unpredicted volcanic eruptions may contribute temporarily to ozone loss as Mt Pinatubo may have done recently. Also, some uncertainties remain in quantifying the chemical processes themselves, although over long periods these errors are likely to be smaller than the errors in representing the atmospheric coupling processes (SORG, 1996).

2.3. Ozone changes over the UK since the late 1970s.

The combination of all the factors noted above means that the rate of change of the ozone column is not constant either from day to day, season to season or year to year and careful measurements need to be made over many years to establish the long-term trend. Ozone depletion has occurred over much of the Earth's surface as indicated by a number of different measuring systems based on satellites and on the ground (WMO 1995, p.1-19). At latitude 55°N, the trends in the period 1979-1991 averaged over all longitudes, as shown in Table 2.1.

The magnitude of the change varies somewhat between the different instruments but it is clear that ozone decrease has been observed with all the instruments although the decrease has been steeper during winter and spring than summer and autumn.

A more detailed look at the UK shows that column ozone has been generally falling since the late 1970s, in common with other mid-latitude regions. The Meteorological Office record for Southern England (Bracknell and Camborne) using a ground-based instrument indicates that column ozone has decreased by on average 4.3 % per decade for the period 1979-95. As with the data in Table 2.1, the record shows a distinct seasonal difference with winter/spring (December - May) decreases of 6.4% per decade and summer/autumn (June - November) decreases of 2.0% per decade. Hence, during the 16 years of the record, column ozone has decreased

Table 2.1

Ozone trends at 55°N (% per decade)

	Satellite: Average of several instruments	Ground-based	Mean
Annual	-3.8	-3.8	-3.8
Dec-Jan-Feb	-3.9	-4.8	-4.2
Mar-Apr-May	-4.6	-4.6	-4.6
Jun-Jul-Aug	-3.2	-3.7	-3.3
Sep-Oct-Nov	-3.4	-1.9	-2.9

about 10% during winter/spring and 3% in summer/autumn. At Lerwick, Shetland Isles (60.13°N, 1.18°W), measurements have been made for several decades but early data were not well calibrated. Starting from February 1981, the mean ozone depletion rate was 4.2% per decade. Again, the trend has been greater during winter/spring with a mean decrease of 6.2% per decade while during summer/autumn the measured decrease has been 3.1%. These results are again in broad agreement with the observations in Table 2.1. Finally, there is also a suggestion that the increase in the downward trend in ozone over the UK for the period 1981-93 (Farmer, 1993) was larger than during the period 1981- 95.

In the winters of 1991-92 and 1992-93 unusually low ozone values were recorded over northern Europe. The precise cause of this is unclear but has been partially attributed to the natural effects of weather systems in redistributing ozone, coupled with an enhancement of the depletion by chlorine chemistry. It is also probable that the eruption of Mt. Pinatubo contributed to ozone loss by providing small liquid droplets (aerosol) in the lower stratosphere on which additional chlorine and bromine based chemistry can proceed. This would explain the increased downward trend up until 1993 noted above. These aerosol effects have now receded and are not therefore indicative of an increased long-term trend, but under some atmospheric conditions ozone will continue to be vulnerable while chlorine and bromine levels remain high. There is now strong observational evidence for significant chemical destruction of ozone during recent winters in the northern high latitudes. During winter 1995/96 the polar stratosphere was again very cold at high latitudes and in early March 1996 record low ozone amounts were recorded at both Camborne and Lerwick. Nonetheless the ozone amount recovered soon afterwards and the monthly means were more consistent with the long term ozone trends.

2.4. UV measurement

The most frequently measured quantity of solar UV radiation is termed global radiation, that is the radiant energy falling upon a horizontal surface from all directions. It consists of two components: direct and diffuse (scattered) radiation. There are many factors which affect the incident global radiation, and the proportions of direct and diffuse radiation. The major determinants are listed in Box 3.

There are three main types of detector for solar UV measurements:

- spectroradiometric monitoring equipment (which measures solar UV at each wavelength interval over a wavelength range generally from ~290 to 400 nm),
- broad-band measurement equipment (providing instantaneous measurement over selected wavelength ranges of biologically weighted irradiance or filtered flux), and
- personal UV dosimeters (generally small biological or chemical detectors to provide the integrated weighted UV dose at various body locations). Personal dosimetry often measures the radiation incident on surfaces of different orientations eg the chest or shoulder. An example of such a material is given in Chapter 8.

The characteristics of the detector used and its calibration are important in determining the measurement accuracy and the uses to which the data can be put. Desirable features in any detecting system include stability, adequate sensitivity, low background response, plus known and reproducible wavelength, angular response and dose characteristics.

There are three main problems related to the collation and comparison of data from these measurements:

- lack of standardisation of instrumentation. Instruments with different characteristics and precision have been used at independent locations, making any comparison of the results from different organisations difficult,
- inconsistencies in the cross-calibration of measurement equipment used by different organisations at various sites. A variety of calibration sources with different characteristics have been used, and
- lack of continuity of measurement and incompatibility of data acquisition systems and data formats.

Most commercially available UV instruments measure global erythemally weighted irradiance (see Box 1, Chapter 1) (typically to a precision of 10-20%), which may be adequate to assess most of the damaging effects of UV on biological systems.

Figure 2.1a

Two spectra measured at Reading on 10/3/95 at 1200 and 23/3/95 at 1400. The solar elevation was 34° in both cases. On the 10th column ozone was 385 DU, and on the 23rd it was 285 DU.

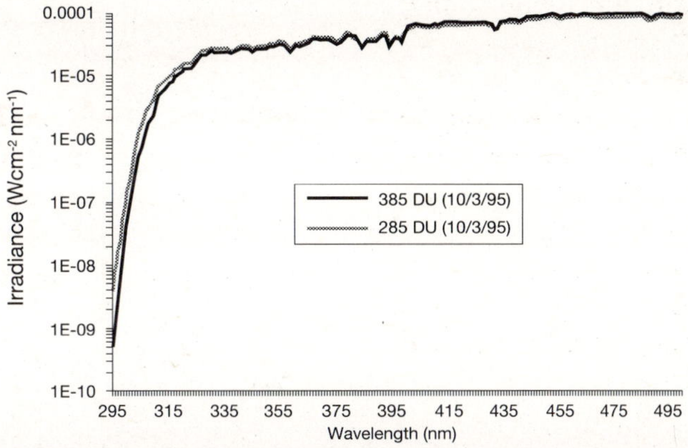

Figure 2.1b

The ratio of the two spectra in Figure (a), 23rd: 10th. The flat line is the ratio of a visible photodiode sampled throughout the two scans. The consistency of this value indicates stable conditions during both scans, and also indicates that solar irradiance on the 10th was in general slightly higher than on the 23rd (also seen at long UVA lengths, unaffected by ozone, in Figure 2.1a).

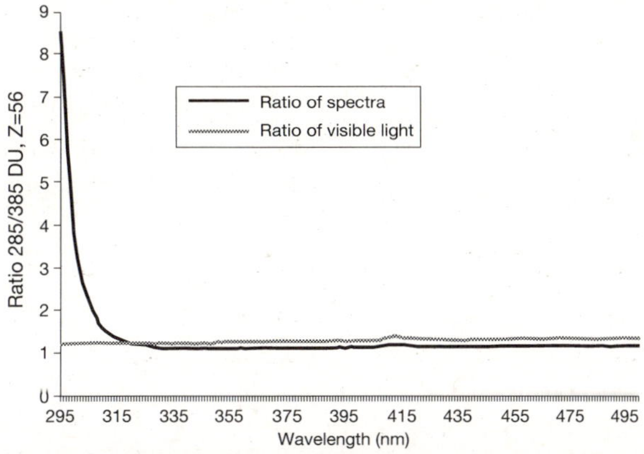

Box 3

The amount and spectral distribution of solar UV flux (irradiance) reaching a detector at the Earth's surface depend on a number of factors, including

(a) the wavelength of the radiation

(b) the solar elevation, which depends on latitude, date of the year and time of day

(c) the solar spectrum incident at the top of the atmosphere

(d) the ozone column thickness

(e) generalised molecular and aerosol absorption and scattering e.g. Rayleigh scattering

(f) absorption, scattering and reflection by clouds

(g) localised atmospheric pollution

(h) the reflectance characteristics of the ground and surrounding environment

(i) the location, instrumental and geometric characteristics of the measuring detector

(j) shadowing of the detector by surrounding objects

(k) the altitude of the detector above sea level

Figure 2.2

Calculated spectral irradiance for 20 June at 55°N under clear skies. Panel (a) shows the irradiance at the top of the atmosphere and the surface irradiances for a range of column ozone values from 216 to 540 DU. Panel (b) shows the part of the spectrum that is sensitive to ozone amount. The top, solid line in each panel shows the values observed at the top of the atmosphere; the lower, solid lines show the value for 360 DU column ozone.

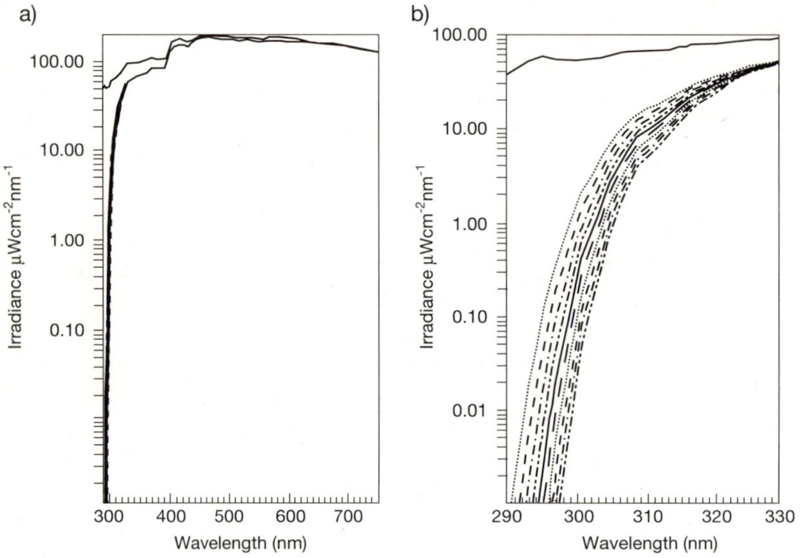

However, atmospheric modelling also requires a knowledge of the spectral irradiance in order to adequately predict the environmental and health impacts of ozone depletion. High precision spectroradiometric measurements of solar UV are required (particularly in the wavelength range 290 - 315 nm) to detect the predicted small trends in UV often indicated by stratospheric ozone modelling and satellite data.

Although there is no clear evidence of a long-term trend in overall UV levels reaching the UK, there are good examples to show that when ozone amounts are low, UV irradiances are correspondingly high. This can only be fully illustrated when the sky is cloudless and the attenuation of UV by cloud does not have to be considered. Comparison of measurements made at the same solar elevation and different ozone amounts will then show the effect of ozone on UV at the surface, all else being equal (factors a,c,e and g-k in Box 3). One such comparison is shown in Figure 2.1a for spectral measurements made at Reading on 10 March and 23 March 1995 at the same solar elevation (34°) for column ozone of 385 DU and 285 DU. The ratio of the two measurements increases dramatically as wavelength decreases across the UVB part of the spectrum (Figure 2.1b). This is a characteristic signature of ozone absorption and reinforces the message that low ozone means high UV. In the UK, the UV data record is not yet long enough, and ozone changes not severe enough, to identify a long-term trend in UV against the year to year variation in cloud amount.

A calculation of the effect of ozone change on UV irradiance at 55°N is shown in Figure 2.2 which may be compared with Figure 2.1a but over the full observed variation of ozone. The first panel shows the results over a wide spectral range while the second panel shows the results over the wavelength range 290 - 330 nm which is most influenced by the ozone amount. The solid lines at the top of the figures are the irradiances at the top of the atmosphere while the other lines give the irradiances at the ground for noon mid-summer. These lines show, from left to right, the range of results for ozone varying from 216 to 540 DU in steps of 36 DU with the solid line in the centre of the figure showing the 360 DU column ozone results.

Periods when ozone is low and skies are cloud free are generally the times when UV levels will be highest for a given time of year and protection against the sun may be appropriate. To aid the public in this respect NRPB issue weekly retrospective bulletins of incident UV (Driscoll and Rawlinson 1995) as an educational tool (Figure 2.3), and the Meteorological Office offer UV forecasts during the summer months.

Figure 2.3

Weekly UV report for Chiltern.

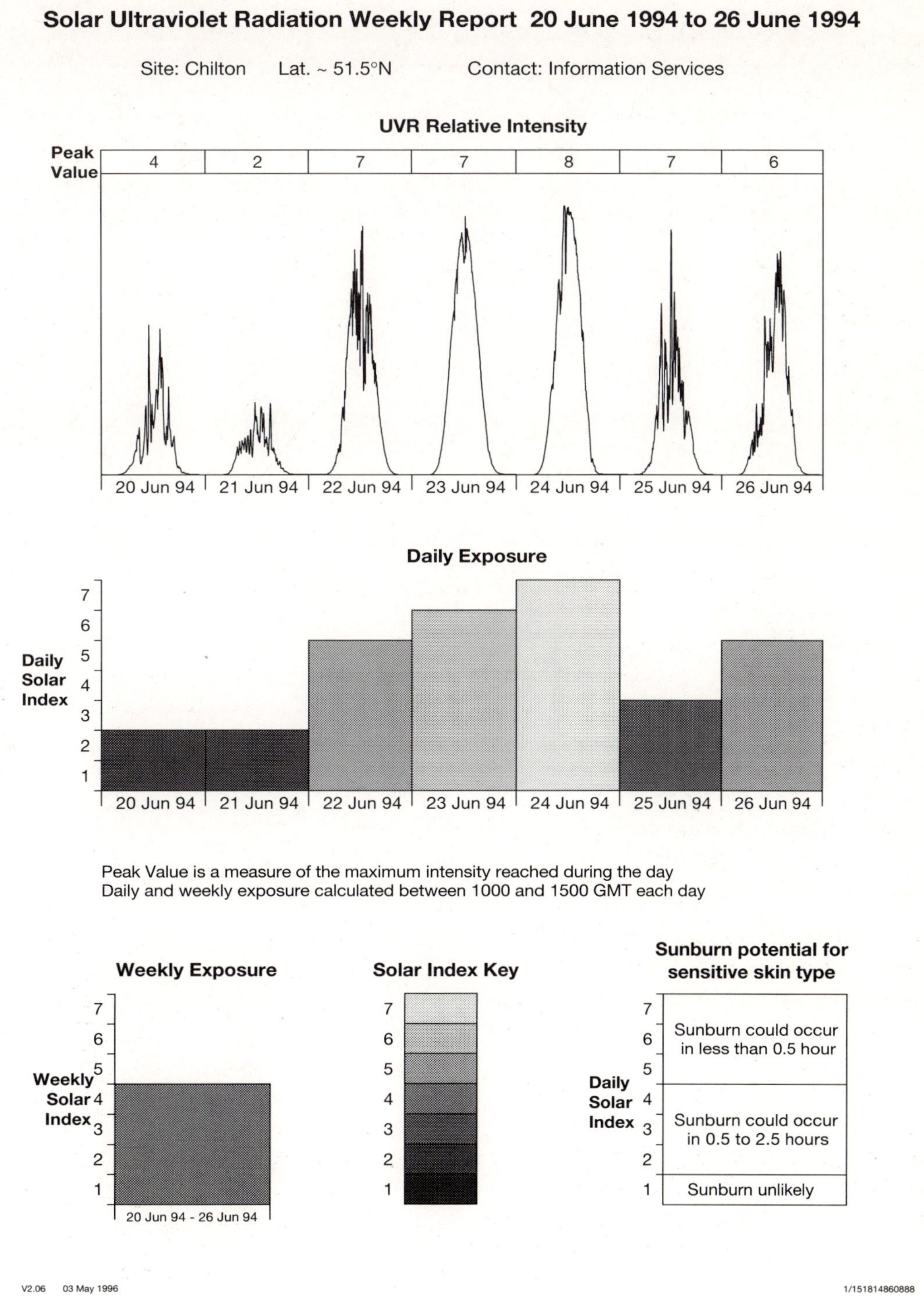

2.5. Forecasting of surface UV.

A number of countries have now established services to forecast UV levels or to inform the public about the health consequences of high levels of exposure. Most of these forecasting and advisory services apply to sunny conditions since cloud can significantly reduce the UV amount. In Europe, a forecast service has recently been established in Finland because of its close proximity to the Arctic where ozone depletion is likely to be most marked. Sweden and Denmark are also developing UV information services. While southern European countries receive much higher UV levels, the perceived risk to their population is lower because their skin types generally afford better protection. Nonetheless, northern Europeans travelling to southern Europe, e.g. for summer holidays, need to take special care. In the UK, the Meteorological Office has been forecasting UV since 1994 (Austin *et al.,* 1994) based on the Canadian method (Burrows *et al.,* 1994). The service was designed to provide warning during sunny conditions in summer, regardless of ozone levels, but future ozone depletion may make the forecasts more pertinent.

The UV forecast method uses several procedures: first the column ozone must be predicted, followed by the UV which can then be presented in a number of ways (see Box 4). The largest day to day variations of column ozone come from its movement from one region to another, while chemistry acts over periods of several months. Hence, forecasting column ozone can be achieved satisfactorily from the meteorological fields. In the UK, the Meteorological Office derives ozone observations from satellite data and these are used to determine regression coefficients relating the column ozone to meteorological analyses in the lower stratosphere. For southern England, the forecasts have been compared with spectral measurements made at Reading University (section 2.4). This suggests that the UV index error is about 7%, which is about a third or a quarter of what would be obtained using climatological rather than forecast ozone. Improvement in the UV forecast accuracy might be achieved through a more sophisticated ozone forecast but the wide range of sunburn times for a given UV index suggests that this would not be justified. Nonetheless, by forecasting the ozone column using a simple statistical method, the conditions likely to yield increased UV can be monitored and suitable warnings issued. These forecasts only address short term maxima in UV exposure (acute exposure). For some UV impacts a slight (climatological) increase over a prolonged period (chronic increase) can be more important (see impact Chapters).

Box 4

The UV forecast method consists of the following steps:

1. An ozone forecast is produced by applying regression coefficients to the forecast meteorological fields.
2. Established empirical relationships are then used to forecast the UV levels under clear skies.
3. The UV can then be presented as a UV index or converted to a sunburn time for dissemination to the public using physiological data (Pathak & Fanselow, 1983) assuming sensitive skin.

A measure of UV is obtained by weighting the wavelength-dependent flux by the erythemal action spectrum (McKinlay & Diffey, 1987), giving values up to about 300 mW/m^2 over the Earth's surface. It is becoming standard to specify the UV in terms of a UV index obtained by dividing the flux by 25 mW/m^2. Figure 2.4(a) shows the clear sky UV index at noon as a function of day of the year at 55 °N. The middle curve is the value calculated from climatological ozone and the outer curves show the range of extremes within a 10° latitude band centred at 55° N for 11 years of the Total Ozone Mapping Spectrometer satellite data record 1981-1991. Days are numbered from 1 January. The UV levels as a function of time of day are given in Fig. 2.4(b) for the solstices (solid lines) and equinoxes (dashed lines) for an open space providing no shade. Lower UV levels are generally experienced during the Spring equinox when column ozone is higher than during the Autumn equinox. UV levels are higher at increased altitudes because of reflections from snow surface and due to reduced atmospheric absorption and scattering. Only limited studies have so far quantified these effects (e.g. Blumthaler *et al.,* 1994).

Figure 2.4

a) Shows the calculated UV index for clear skies at 55°N throughout the year. The centre line indicates the values expected at noon for clear skies assuming climatological ozone column. The outer curves show the values which would have been expected assuming the extreme ozone values recorded by the Total Ozone Mapping Spectrometer instrument near 55°N during the period 1981-1991.

b) Shows the calculated UV Index for clear skies as a function of time of day for the solstices (solid lines) and equinoxes (dashed lines)

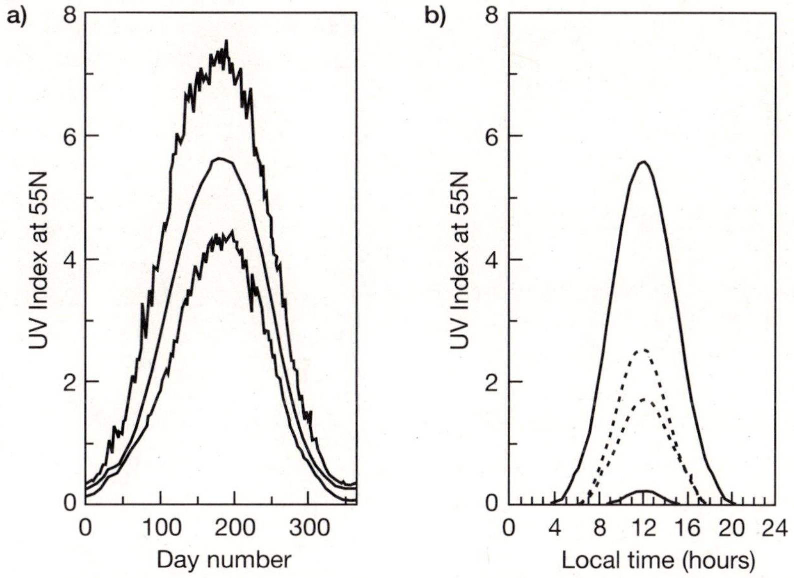

2.6. Globally observed changes in UV

There are no unbroken records of UV measurements which cover a true climatological time scale (30 years plus), and with no changes to the instrument or measurement procedures. This makes it difficult to identify long-term trends in UV radiation at the Earth's surface. In addition, UV at the surface depends on other features of the atmosphere besides ozone, so its natural variability is a complex mixture of the cycles and changes in ozone, cloud, meteorology and pollution which may offset or enhance one another at any given time. Determining a trend in UV is thus more difficult than determining a trend in ozone. It is important to recognise the difference between a continuing pattern of change underlying a varying signal (a trend), and extreme events which are changes from the short-term norm but may be expected on a climatological time scale.

The longest UV records come from the USA, Russia and Switzerland. A network of broadband instruments in USA reported a slight decrease in UV between 1974 and 1985, and this was matched by data from Moscow for approximately the same period. Subsequent investigations and reworking of the US data now suggests that the UV may have increased slightly during the period in question. A multi-filter instrument in Maryland recorded increasing UV from 1974 until 1987, at which time changes were made to the instrument and no further increase was observed. Discrete periods of broadband measurements at high altitude in the Swiss Alps since 1982 have been combined to indicate a small statistically significant increase in UV which is in keeping with the reported ozone trends. The results are discussed in more detail in WMO (1995).

The increase in UV monitoring in the 1990s means that short-term changes in UV have been observed at a number of locations and correlated with low ozone events. The most dramatic increases in surface UVB have been seen in Antarctica during the ozone hole events in the springtime. When the ozone hole extends over southern Argentina similarly elevated UV is observed, indeed the absolute levels of UV can be much higher here than in the Antarctic because of the lower latitude. Despite an underlying indication of a persistent upward trend in UVB, the yearly maximum values and the variability in monthly dose depends as much on cloud as on ozone. As the Antarctic ozone hole

Figure 2.5

Map showing NRPB Measurement Sites, with UK weather (March 1-6, 1996) and peak UV and low ozone values at the sites

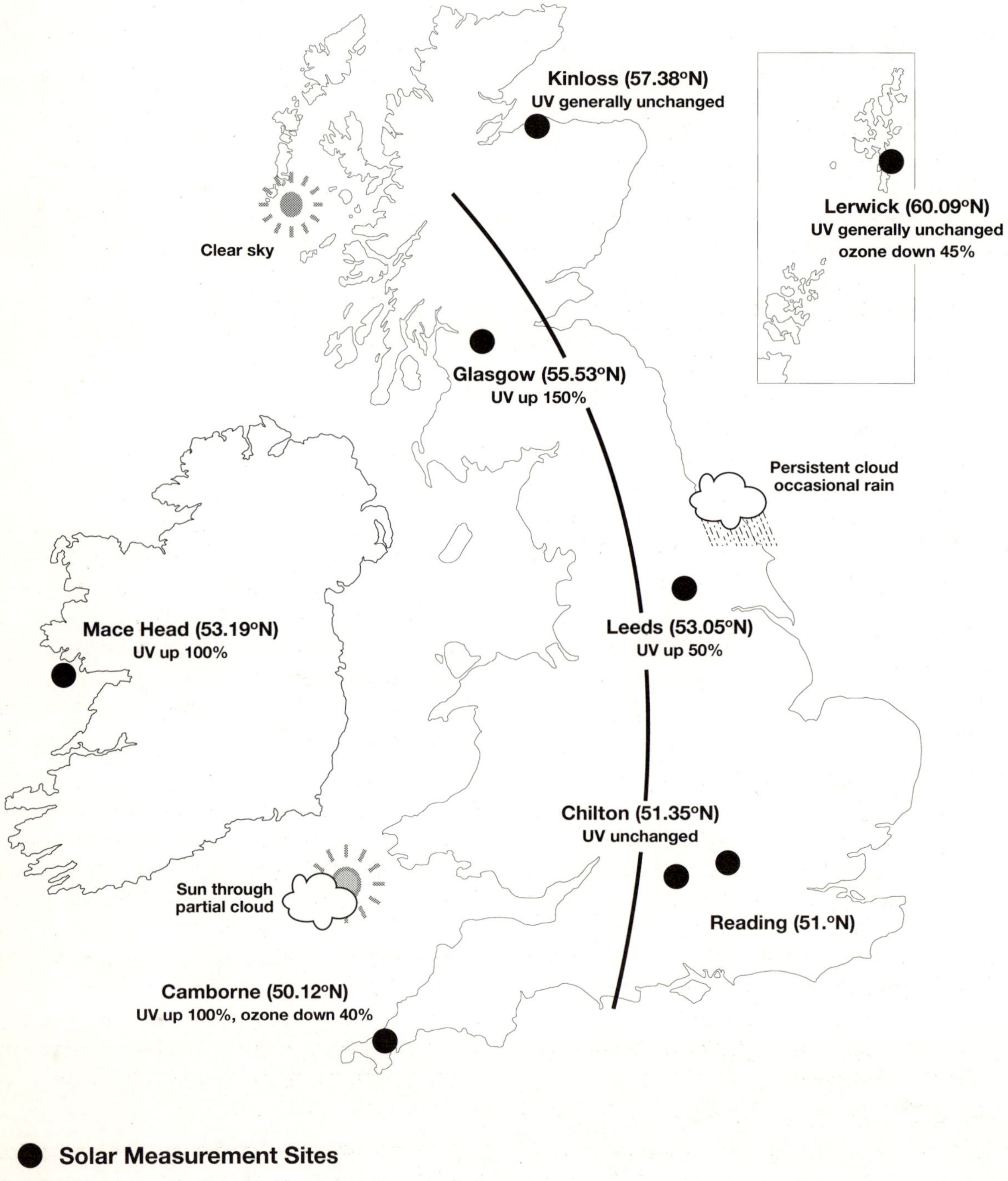

breaks up pockets of low-ozone air have been seen to drift over southern Australia, giving associated high UV levels when coinciding with clear skies. Elsewhere, seasonally elevated UV levels have been reported over Canada, Germany, Austria, Scandinavia and Alaska during 1992 and 1993. Then, aerosols from Mount Pinatubo in the stratosphere plus an unusually persistent meteorological situation combined to give noticeable and persistently low ozone through much of the winter and spring, and through into the summer months. Such obvious excursions from the norm have not been observed since, although there were a few periods of low ozone over the Arctic regions in winter/spring 1995-96 associated with a short-lived meteorological feature. During these recent low ozone events corresponding high UV was observed in Scandinavia and Northern Europe. Low ozone, clear skies and high UV also occurred over the UK for several days during the same months, but can be attributed more to a strong anticyclone than to ozone depletion. It must be stressed that the absolute levels of UV at mid - to high latitudes during the winter months are low even when there is reduced column ozone, and absolute changes in UV will be much more significant if low ozone occurs in summer, as in 1993. During the early months of that year no increase in broadband measurements of UV dose could be identified because of the highly variable cloud.

Figure 2.6

Average monthly values of daily spectral irradiance at 310nm and 55° N. These are calculated for clear skies and observed ozone values, and expressed as deviations from the corresponding calculated values for 1978-1993 average ozone. The thick lines and left scale gives absolute irradiance changes, the thin line and right scale gives percent changes (reproduced from WMO/UNEP 1994/5).

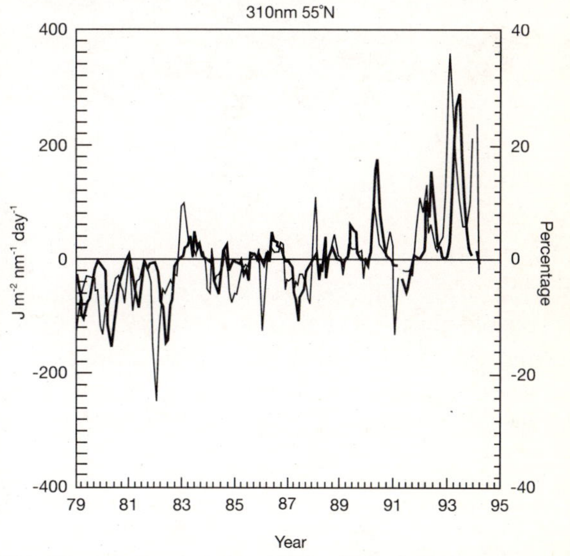

2.7. Measurements in the UK

During the 1970s and 1980s, sporadic measurements of solar UV levels were made at various locations in the UK. Early measurements were made by the Building Research Establishment using a multifilter instrument with one band in the UVB and two in the UVA, but little data are available. Spectral measurements of radiation were made from a reflecting plate at the University of Surrey in 1974, and then in 1982-85 both broadband (continuous) and spectral (clear sky) measurements were made at the University of Nottingham.

The University of Reading has been a major contributor to spectral UV measurements with measurements which date back to late 1989, and a near continuous record since 1993. The early reported spectral measurements often relate to clear or near clear sky conditions (Webb, 1991), which are limited to less than ~15% of the total days throughout the year over the UK. The all-weather monitoring therefore gives vast improvement in the coverage although measurements in cloudy conditions are less straightforward to interpret. The instrument at University of Reading has participated in several international intercomparison campaigns which have verified the instrument and operational procedures in use, and provide a link to similar measurements made elsewhere in Europe and the world.

Broad-band measurement data, often inherently weighted with the Commission Internationale De L'Éclairage (CIE) erythemal action spectrum (McKinlay and Diffey, 1987), have been collected by workers based in hospitals in Durham, Glasgow and Dundee. However, many of these data have not been published, being primarily for use in in-house research. Broad-band measurements, made instantaneously and continuously, may be used to monitor short-term changes in sky and environmental conditions.

In 1988, the National Radiological Protection Board (NRPB) set up three monitoring stations to make continuous measurements of solar radiation at different latitudes within the UK (Fig. 2.5). These are at the NRPB centres at Chilton (latitude ~52°N), Leeds (~54°N) and Glasgow (~56°N). Measurements of visible (approximate wavelength range 400 - 770 nm), ultraviolet A (~320 - 400 nm) and erythemally weighted ultraviolet radiation (~290 - 400 nm, UV_{eff}) have been made simultaneously and continuously using a system based on three commercially available broad-band detectors. The measurements are designed to provide information regarding the range of variation of solar UV at different latitudes in the UK with the time of year and from year to year, and to establish baseline levels for natural UV. Readings are recorded every 20 seconds and these are then used to produce a set of mean values every 5 minutes, together with associated statistical deviations.

During 1993-94, the NRPB measurement network was extended to cover the range of latitudes (from ~50°N to ~60°N at every ~2° of latitude) appropriate to the whole of the UK (Fig. 2.5). This was achieved by establishing measurement sites at Meteorological Office locations at Camborne (~50°N), Kinloss (~58°N) and Lerwick (~60°N). In addition, spectroradiometric measurements have commenced on a continual basis at the NRPB sites at Chilton and Glasgow.

Linear regression analysis of the annual integrated solar UV_{eff} measurements at Chilton for the 7 complete years from 1989 to 1995 indicates no significant change in the annual integrated UV_{eff} values at Chilton over the measurement period, although this is considered to be a relatively short period for statistical analysis (Lubin and Jensen, 1995). A more selective analysis for trends in UV data as a possible consequence of ozone depletion is required for clear sky conditions (Blumthaler *et al.,* 1994) . However, the data set is severely reduced in such an analysis since on average less than 15% of the days in the UK relate to clear or near clear sky conditions. Although the data record is too short to show significant changes in UV, the expected changes in UV for clear skies due to ozone variations 1978-93 have been calculated for 55°N and are shown in Fig. 2.6. Although the absolute values of change in this figure are greater than have been received (because much of the time it was cloudy), the percentage changes should be valid assuming that the weather is independent of any change in ozone. Since 1992, the percentage UV increase above the long term mean has been positive.

During early March 1996 (Fig. 2.5), at the same time that record low levels of ozone were recorded over the UK, the erythemally effective UV levels at some of the NRPB network monitoring sites increased above expected levels (Driscoll, 1996). The levels at Camborne, Glasgow and Mace Head were double those normally observed during early March, while transient levels at some of the other sites also increased. The greatest impact appeared on 3 March, with large increases in both the peak and daily UV_{eff} levels for west coast sites. The western part of the UK and much of Ireland had weather conditions during early March that allowed the sun to appear through broken clouds, while eastern, northern and central parts had more persistent cloud and occasional rain. Therefore, the local weather dictated where increased UV occurred. Although the event was short lived and did not represent a significant health hazard, it is the first occasion on which UV_{eff} levels substantially higher than normal have been measured simultaneously at several sites across the network at a time of low ozone.

2.8. Future developments in UV measurement

Future developments of UV radiation measurement are likely to concentrate on:

- The coordination of existing spectral and broad-band radiometers at various locations to provide a standardised framework for measuring solar UV as a function of geographical location, time of day and year and atmospheric conditions.
- The establishment of quality assurance procedures in solar radiation measurement including a reliable and inexpensive method of cross-calibration of measurement equipment at the sites by providing, for example, a series of travelling calibrated sources and radiometers to be sent sequentially to all measurement sites.
- The correlation of solar radiation measurements with measured and published data relating to stratospheric and tropospheric conditions relevant to the UK, e.g. aerosols, cloud thickness and type. Few of these parameters are measured on a routine basis, but if they were to become available it would increase the value of the UV data in interpreting atmospheric conditions.
- The establishment of a computer based data storage and retrieval system for UV and for atmospheric and stratospheric data

relevant to selected measurement locations so that the data can be collated centrally on a regular basis to examine spatial and temporal distributions and trends.

The World Meteorological Organisation has recently established a scientific steering committee to address this issue globally.

2.9. Likely changes in ozone and UV over the next 20 years.

Future ozone predictions are made largely on the basis of simulations with mathematical models of the atmosphere which contain a description of the chemistry affecting ozone and the meteorology controlling its movement by stratospheric winds. Because of their computational expense, such models are usually highly simplified and in particular often represent the atmosphere as two-dimensional (latitude-height) with symmetry in the longitudinal direction. This can be a particular problem in the representation of those reactions which only occur if the local atmospheric conditions are suitable. Nonetheless, broad agreement has been obtained between model results and observations of column ozone over the period 1980 - 1990, providing some confidence in the models. However, none of the models in the recent model intercomparison (WMO, 1995) reproduced the 5-10% per decade decrease in ozone concentration in the midlatitude lower stratosphere observed from satellite data (McCormick *et al.,* 1992).

Despite their shortcomings, results from two-dimensional model predictions are used here to estimate future conditions according to two scenarios, I and II. In scenario I chlorine and bromine concentrations and greenhouse gases are in accordance with the Montreal Protocol and amendments. In scenario II only partial compliance is assumed as indicated in Table 6.4 of WMO (1995). In both scenarios, 2-D model results suggest that at 55°N, the column ozone will reach a minimum in 1996-1998 at only 1-2 % below 1995 values, i.e. about 11-12% below 1980 values for winter/spring. In the case of scenario I, by about 2010, the ozone depletion will have reduced by half while in scenario II, the ozone recovery could only be 1-2 %. Figure 2.7 shows the effects of these changes on erythemal UV at midday throughout the year. The lower curve shows the UV index for 1980 at 55°N, based on satellite data. The upper curve shows the UV index at the time of minimum projected ozone under scenario I and the middle curve shows the UV index in the year 2010 as predicted by the 2-D models under scenario I. Note that the predicted UV change is small compared with the day to day variation of about 30% (see Fig. 2.4) for clear sky conditions. A systematic ozone decrease would nonetheless lead to an overall increase in UV dose.

More complex models of the atmosphere which treat the chemistry in three dimensions are now becoming available and are being used to investigate the combined effects of increases in chlorine and in greenhouse gases. In seasonal integrations of such a model for a future doubled CO_2 atmosphere, Austin *et al.* (1992) simulated severe Arctic ozone depletion. However, the effects in any given winter are highly dependent on the characteristics of the weather systems in the lower atmosphere. Rind *et al.* (1990) for example, found

Figure 2.7

Calculated UV index at 55°N throughout the year. Lowest curve: values assuming 1980 column ozone values. Upper curve: projected values expected from 2-D model ozone calculations at the time of the minimum ozone levels. Middle curve: projected values expected from 2-D model ozone calculations at about the year 2010 assuming compliance with the Montreal protocol.

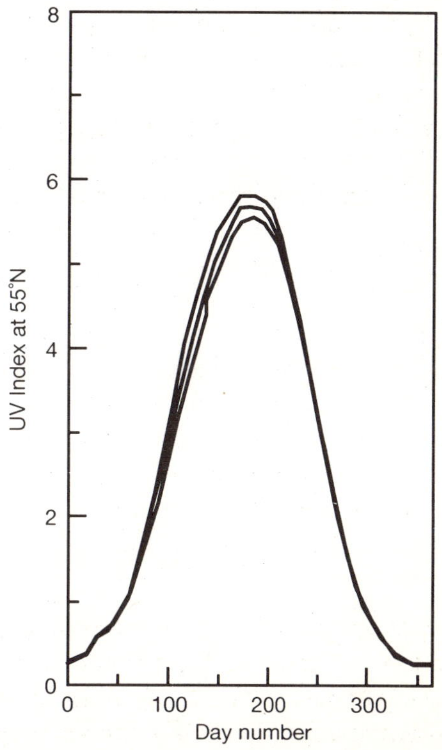

that increasing CO_2 in general produced a more disturbed lower atmosphere which would tend to make ozone depletion less likely. The calculations of Austin *et al.* (1992) suggest local ozone depletions of about 50% may be possible in late spring at 80°N but their calculations gave less than 10% depletion at 55°N due to doubling CO_2. The calculations were not extended to summer because the model oversimplified the effects of the lower atmosphere which prevented long simulations. The observation of severe ozone depletion in several of the most recent winters suggest that it is possible that some chemistry/climate coupling effects are already present. Several more years' measurements will be required before we can be certain that the most recent winters are not statistical deviations.

Natural processes such as the recent eruption of Mt. Pinatubo can also contribute to ozone loss. In this case the volcanic aerosol acts as a site to transform inactive chlorine into active forms which can then destroy ozone. Thus, the presence of anthropogenic chlorine has the effect of making stratospheric ozone more vulnerable to year to year variations in volcanic eruptions and weather systems. We could therefore experience far more future variability in ozone amounts which are not represented by current two-dimensional models. There is therefore a need to perform three-dimensional model calculations incorporating more completely the effects of the tropospheric source gases on stratospheric ozone and stratospheric temperature.

3. Molecular Effects

- **Ultraviolet radiation interacts with specific molecules within the cell to produce photodamage.**
- **Photodamage yield and nature are dependent upon wavelength.**
- **Photodamage may be modified by repair processes.**
- **Photodamage generates lethal and mutagenic events leading to cancer.**

Basic Processes

To influence living things light and its component UV wavelengths must be absorbed by a set of molecules (*chromophores)* in order to lead to chemical and physical changes in the cell. The first step in this photochemical reaction is the absorption of energy and the generation of an excited state in the absorbing molecule. The products generated by this process are either reactive species in an excited state or free radicals. Both of these products are formed extremely rapidly and this is followed by reactions which generally occur within microseconds but which may last for some hours. These rapid processes constitute the basis for photobiological responses.

Damage to Biological Molecules

Since the chromophores which interact with ultraviolet radiation are defined by the wavelengths at which they absorb, the nature of the critical chromophores is a function of wavelength throughout the whole UV spectrum. Thus the peak absorption of deoxyribonucleic acid (DNA) occurs at around 260nm, this falls off through the UVB region and becomes effectively undetectable in excess of 320nm (see Fig. 1.1, Chapter 1). Protein absorption peaks in the UVC region but also extends into the UVA region in the form of amino-acids such as tryptophan indicating that direct damage to proteins can occur at much longer wavelengths than to DNA. UVA chromophores present in the cell are represented by molecules such as quinones and porphyrins.

Many cellular targets including nucleic acids (DNA and RNA), proteins and lipids can be damaged by UV radiation. With wavelengths in the non-solar UVC region, DNA is the most important cellular target and the majority of these changes are the consequence of direct action. Solar UVB also damages DNA by direct action but the extent of oxidative damage in this molecule remains controversial. At wavelengths in the UVA region any damage to DNA is likely to be the consequence of the action of active oxygen intermediates generated by photosensitising agents such as flavin containing compounds which can occur naturally within the cell or by chemicals which may be taken incidently in the diet or as a medicine.

Exposure of DNA to solar UV radiation results in a range of chemical changes (photoproducts) which have been the subject of extensive laboratory investigation and the ratio of these different photoproducts varies considerably with wavelength. One, the cyclobutane type pyrimidine dimer (normally referred to as the "dimer"), occurs in greatest yield and as a result has been studied in most detail (Setlow, 1966) see Fig. 3.1. Other chemical changes which are induced in lower yield include the pyrimidine (6-4) pyrimidon photoproduct and its "dewar" photoisomer (IARC 1992).

With regard to ribonucleic acid, the rapid turnover of messenger RNA suggests that these molecules are not critical in cellular response.

Damage to proteins becomes more important as the wavelengths increase in the UVA and UVB regions because of the increased absorption of aromatic amino acids in comparison with nucleic acids.

Protection and Repair

Both plants and animals are able to generate photoprotective mechanisms. In plants the production of flavonoids is increased in response to UV. These compounds function as radical scavengers like glutathione in mammalian cells (Tyrrell and Pidoux, 1988). Other naturally occurring radical scavengers include vitamins C and D. Tanning, which reflects the formation of melanin, modifies the sensitivity of skin and taken together with skin thickening represents a protective mechanism for humans.

Any damage to DNA which arises despite the intervention of protective processes can be reversed by repair which may or may not make errors and thus lead to mutations (Freidberg *et al*., 1995). The two most important enzymatic DNA repair processes are photoreactivation and nucleotide excision repair. In photoreativation a single enzyme complex (photolyase) is activated by short

Figure 3.1

Action specta for Cyclobutane pyrimidine dimer formation in naked DNA and mammalian cells grown in tissue culture

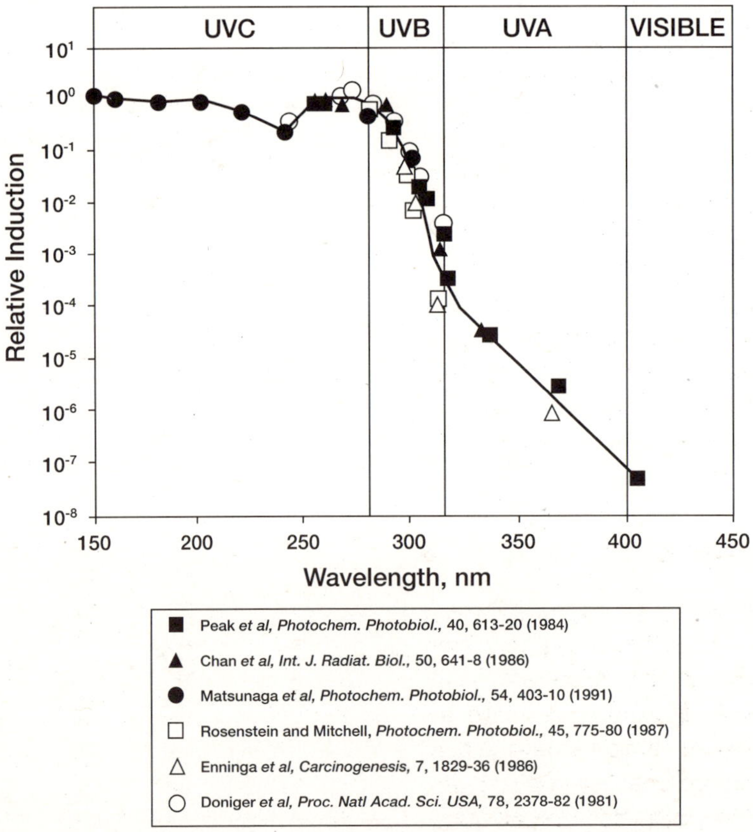

Source: Board Statement on Effects of Ultraviolet Radiation on Human Health, Vol 6, No 2 (1995) courtesy National Radiological Protection Board

wavelength visible light and results in the direct splitting of the cyclobutane ring between two adjacent pyrimidines generated by absorption of UV radiation. This activity is generally held to be largely error free and plants are particularly efficient in this photoreversal process. Amongst mammals only the monotremes show functional photolyase activity. Nucleotide excision repair is brought about by a complex of repair enzymes including transcription factors whereby the damage is recognised, cut out of the DNA, a new stretch is synthesised using the complementary strand as a template and then patched in to restore the integrity of the original DNA sequence (Bootsma and Hoeijmakers, 1993).

Cellular Targets

For UV radiation DNA is taken as the most important cellular target because of its unique genetic status. UVB is absorbed by DNA and a variety of photochemical damage is produced implicating DNA as the primary cellular chromophore and site of damage for most of the biological consequences of exposure to shorter wavelengths of UV. The damage induced by the UVB component of sunlight in DNA leads to the generation of mutations. At longer wavelengths the targets will be different, for example, the destruction of mitochondria and membranes will become more significant the greater the wavelength.

In plants, in addition to damage to DNA, increased exposure to UVB may alter the basic processes of photosynthesis through changes in proteins including enzymes and pigments (Bornman and Teramura, 1993). Modifications to plant growth form result from molecular mechanisms which remain poorly defined but which may include UVB specific photoreceptors and growth regulators (Caldwell *et al.*, 1995).

Biological outcomes

The cellular consequences of UV damage may conveniently be assigned to lethality or mutation. Studies with cells growing in tissue culture have shown that the majority of mutations arise at dipyrimidine sites further emphasising the importance of pyrimidine lesions (Mitchell *et al.*, 1993). For animals and in particular humans the outcome of mutation is skin cancer. Cancer is defined as a multistep process. The first, initiation, step is a viable mutation and analysis of the mutations which occur in human and UV induced mouse skin tumours shows that they also occur predominantly at dipyrimidine sites (Kress *et al.*, 1992). The next step(s), tumour expression or promotion, is seen to be dependent upon the immunomodulatory action of UVB (see Chapter 6).

4. Aquatic Effects

- **There is no doubt that exposure to UVB radiation is detrimental to the biota of the aquatic environment.**
- **UVB has its greatest impact either at or immediately below the surface of the water where most primary productivity occurs and the eggs and early life stages of many commercially important species of animals reside for some period of their lives. Tidal and intertidal zones of marine and brackish environments, shallow streams and some fish farms are more vulnerable to impact owing to a limited potential to attenuate UVB radiation.**
- **Anthropogenic contaminants in the water column can photosensitize the biota resulting in enhanced potency of UVB; furthermore, acid rain can mediate a decrease in (UVB attenuating) dissolved organic carbon and increase penetration of UVB in the water column.**
- **The construction of reliable models of effects for use in risk assessments is problematic due to,**
    1. **the absence of a long term data base,**
    2. **a limited understanding of the mechanisms of damage,**
    3. **a lack of knowledge regarding interactions of UVB with diverse environmental variables and**
    4. **limited data for the determination of the transparency of UK waters to UVB.**

4.1 Introduction

Aquatic, mainly marine, habitats cover about 70% of the global surface. As well as its inland waters, direct United Kingdom interest in aquatic ecosystems extends as far as the UK fishing fleet ranges. Fish yields are dependent on marine primary productivity and associated food chains; biological processes that are impacted by diverse anthropogenic (man made) influences most especially in coastal and inland waters.

The UVB radiation climates of aquatic systems differ from those in comparable (sea level, same latitude) terrestrial systems. The attenuation of UVB (specifically 310 nm radiation) by natural waters is always at least twice that of PAR (photosynthetically active radiation) (Kirk, 1994a; b). This results from the wavelength-dependent absorption of radiation by water *per se* and dissolved inorganic and organic solutes, SPM (suspended particulate material), DOM (dissolved organic matter including material variously termed humics, yellow substance, Gelbstoff and gilven and consisting of plant derived, chemically resistant compounds) and living organisms. The range of UVB absorption coefficients (units: m^{-1}) for natural waters world-wide is at least 1000-fold, while that for PAR is over 300-fold.

UVB radiation is only one of a multitude of stressing parameters that can simultaneously, and possibly interactively, impact on the aquatic environment. Examples of other stressors include sewage, acid rain, petroleum derived hydrocarbons and organohalides. As such, any assessment of the likely effect of UVB on the aquatic environment should consider the role that these other factors play in the attenuation of UV, as well as their interacting impact on the biota.

The extent of penetration of UVB in seawater is greater in the open oceans than in coastal waters. Zaneveld (1975) demonstrated that 10% of incident UVB penetrated to around 1 m in coastal waters and down to 15 m in open ocean waters. Furthermore, in clear Antarctic waters Gieskes & Kraay (1990) state that "10% and more of the UVB measured just above the water penetrated to 30 m". Smith *et al.* (1992) recorded UVB radiation at depths in excess of 60-70 m which was described as well below the depths at which effects on primary productivity were observed.

Gelbstoff (see above) is present in both fresh water and sea water, the seasonal variability being higher in the former whilst the latter exhibits greater spatial variability, most particularly in coastal waters (Højerslev, 1988). Aquatic environments thus permit organisms, including phototrophs (organisms which obtain the energy needed for growth from electromagnetic radiation) which need PAR in excess of ~1μmol photon $m^{-2} s^{-1}$ as well as visual predators, to avoid UVB if they can migrate deeper into the water body (what constitutes 'deep' being a function of the UVB and PAR absorption coefficients of the water).

Another difference between the atmospheric and the aquatic environment is the impact of UVB on the chemical environment of the organisms. UVB can, modify the troposphere around terrestrial organisms and the upper mm or so of litter (dead organic matter). However, the concentration of organic matter in natural waters, particularly in the surface microlayer, is greater than in the troposphere, and the mobility of organic matter in natural waters exceeds that in litter. This means that in the upper layers of water bodies UVB has greater potential to cause photochemical modification of biogenic (produced by living matter) and anthropogenic organic matter, and for the photochemical products to then interact with aquatic organisms. Thus whilst some organic compounds in natural waters can help to screen organisms by absorbing UVB, and by increasing availability of essential transition metals, others can exacerbate UVB effects by their increased toxicity due to UVB induced photochemistry (Palenik *et al.*, 1991). Furthermore, the capacity of organic contaminants to act as photosensitizing agents to the biota serves to increase UVB potency (Landrum *et al.*, 1986, Newstead & Giesy, 1987). Although UVB also has the potential to render some contaminants less toxic, and can render natural organic matter more palatable for non-photosynthetic microorganisms (Palenik *et al.*, 1991).

While UVA has effects on aquatic organisms, interacting with UVB (see below), no study seems to have been undertaken on UVA/organic contaminant interactions.

Whilst stratospheric ozone depletion is widely recognised as the principal driving force behind increased UVB exposure, recent studies have demonstrated that acid rain and global warming also contribute to increased UVB exposure in freshwater environments. Schindler *et al.* (1996), studying the consequences of climate warming and lake acidification for UVB penetration, suggest that some of the changes in aquatic ecosystems that have been attributed to acidification may in fact be the result of increased penetration of UVB radiation. Dissolved organic carbon (DOC) is known to attenuate UVB therefore any decrease in DOC, following acidification of the water body, resulting from industrial emissions of the oxides of sulphur and nitrogen, would lead to greater UVB penetration in the water column. Schindler *et al.* (1996) concluded that "in clear, shallow lakes and streams climate warming and/or acidification can be more effective than stratospheric ozone depletion in increasing the exposure of aquatic organisms to biologically effective UVB radiation".

For the marine environment around the UK the make-up of the surface microlayer, including organic contaminants in certain areas, and the density of SPM and Gelbstoff in the water column are in all probability the principle factors governing UVB impact on the biota. Kirk (1994b), states that some 10% of surface UVB irradiance penetrates to a depth of between 3.4-6.2 m in the Fladen Ground of the central northern North Sea, between 0.5-4.3 m in the German Bight and 12.1 m at Shetland-Faroe Islands. According to Højerslev (1988) the level of Gelbstoff around much of the North Sea coast of the British Isles, excluding estuarine plumes where levels are considerably higher, is in the order of half that measured on the Fladen Ground, therefore it must be assumed that UVB penetration will be correspondingly higher. Whilst there is a considerable database related to diverse factors (e.g. Gelbstoff, phytoplankton biomass and DOM) that might effect UVB transparency in British waters, the data appear to be restricted to the southern North Sea, its associated estuaries and the English Channel (Laane & Kramer, 1990, 'NSQSR 1993', Joint & Pomeroy, 1993). However, whilst these may give some indication of the penetration of UVB radiation in the water column they do not necessarily reflect fully on the likely impact on the biota when photochemical modification of contaminants, photosensitisation of the biota and acidification factors are also taken into consideration. Contaminant levels in the North and Irish Seas are well documented (Lohse, 1990, Cofino *et al.*, 1992) and many of the contaminants present include the photosensitizing agents derived from petroleum products including anthracene, phenanthrene, fluoranthene, pyrene and naphthalene.

4.2 Impact on the Biota

Primary Producers: planktonic

The primary producers over much of the UK area of interest are microscopic phytoplanktonic organisms. Phytoplankton are frequently subject to stochastic vertical movements in the upper mixed layer of the water column which means there is little capacity for the organism to maintain a vertical position at a constant distance from the water surface or at a constant ratio of UVB:PAR. Vertical movements as a result of buoyancy regulation or flagella motility are incapable of overcoming all but the most limited (< a few hundred μm per second) vertical water movements. Furthermore, flagella apparatus is particularly sensitive to UVB damage (Häder, 1993) thus restricting UVB avoidance by vertical positioning resulting from flagella motility in habitats with limited vertical water movement. A further aspect of phytoplankton which restricts their ability to avoid UVB damage is their small size which limits the usefulness of soluble intracellular UVB screening compounds in restricting the UVB flux to target molecules. Larger phytoplanktonic organisms (> 10 μm diameter) can, however, have significant UVB screening. The UVB flux to target molecules can be reduced to 20% or less of the value without the screening compounds (Raven, 1991; Garcia-Pichel, 1994). Screening can also be improved by (extracellular) UVB absorbing compounds (Marchant, *et al.*, 1991; Garcia Pichel & Castenhoiz, 1991).

Damage to phytoplankton organisms by UVB has been investigated in terms of growth rate (mainly in laboratory conditions), and the rates of component processes of growth in short-term experiments (e.g. photosynthesis, respiration, uptake of ammonium (NH_4^+) and nitrate (NO_3^-), as well as DNA damage (e.g. Karentz, 1994; Burra *et al.*, 1995; Behrenfeld *et al.,* 1995). Behrenfeld *et al.* (1995, and references therein) argued that effects of UVB on combined N (nitrogen) influx are more relevant to long-term

Figure 4.1

Attenuation of UVB (310nm) with depth in a minimally attenuating seawater and a strongly attenuating freshwater lake
(Note: 100-fold difference in vertical scales)

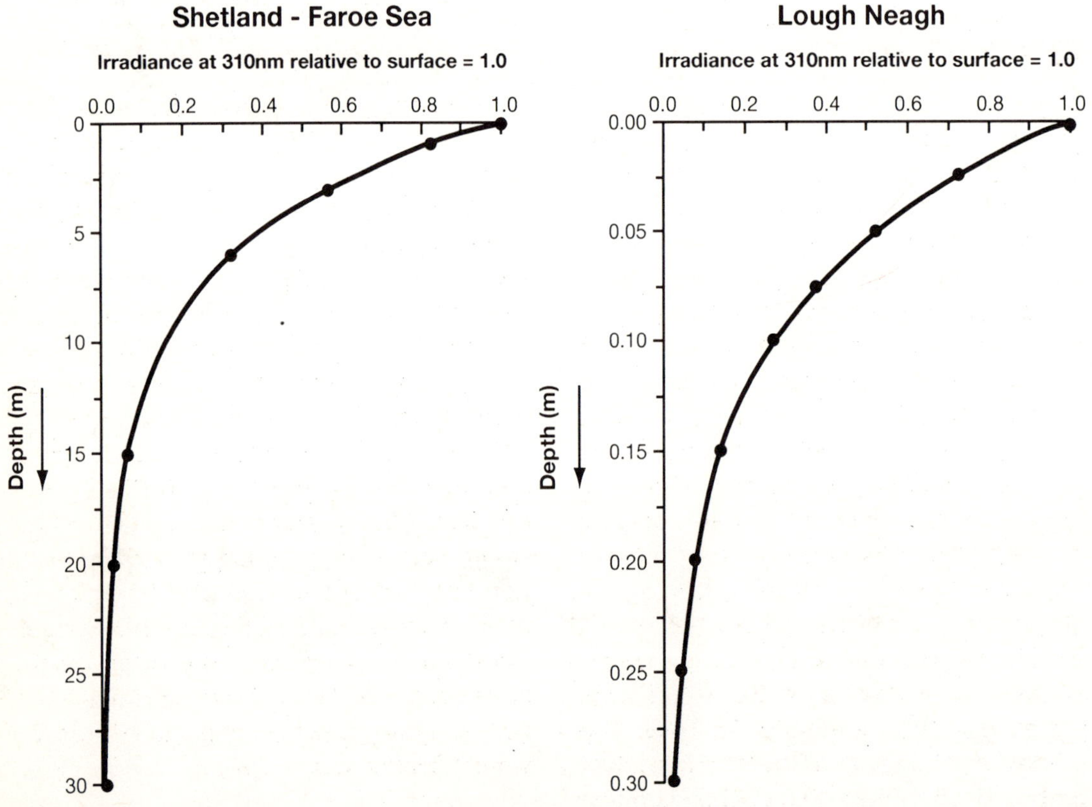

Source: Kirk (1994a)

effects on the primary productivity than the more commonly measured C (carbon) assimilation. This is because N is the nutrient most commonly limiting primary production in the ocean. This work of Behrenfeld and his co workers found that 'realistic' enhancements in UVB above current levels inhibited uptake of combined N under high-N conditions, but that such UVB increases had no effect when growth was already N limited by more than 16%. This means that the process(es) limiting growth rate in low nutrient habitats are not as sensitive to UVB as those limiting the maximum growth rate seen at resource limitation. Perhaps this reflects no more than the most UVB sensitive 'targets' having a lower control strength (Kacser & Burns 1979) in nutrient-limited growth. However, experimentally it is unfortunate that the 'nutrient-replete' data had to come from batch cultures while the nutrient-deficient' data were obtained from chemostats (devices which permit continuous culture of micro-organisms with a known growth limiting nutrient).

A point of some contention is the involvement of UVA in the deleterious effects of UVB on phytoplankton. In some work, a combination of UVA and UVB with a given flux of PAR causes a greater inhibition of photosynthesis than UVA or UVB alone. In other work, UVA can partly offset the inhibition of photosynthesis by UVB (Cullen *et al*., 1992; Smith *et al.,* 1992). Refer Chapter 5.

For the natural environment tidal fronts (that is the outcrop of water occurring at the interface between mixed and stratified waters) represent important physical, chemical and biological boundaries that have been shown to be significant in determining the distribution of phytoplankton (Pingree & Griffiths, 1978). Whilst plankton abundance at fronts is a function of hydrodynamic concentration, the mixing conditions that are responsible for fronts also determine the availability of light and nutrients required for phytoplankton growth. Limited data are currently available for UK waters on the levels of UVB absorbing components in the water column at these frontal boundaries. However, the region of the Flamborough Head Front off the east coast of the UK (Pingree & Griffiths, 1978) has been characterised by Højerslev (1988) as being comparatively low in levels of Gelbstoff thus affording relatively high penetration of UVB. The front lying between Orkney and Shetland has lower levels of Gelbstoff still implying that UVB penetration would be even greater (Fig. 4.1). No information appears to be available for the Gelbstoff levels in the water column at the Ushant, Celtic Sea or Islay-Malin Head Fronts.

A final consideration regarding UVB effects on planktonic primary producers is the sedimentary archive of evidence as to the species composition of phytoplankton in the Antarctic over the years in which the spring ozone hole has developed. McMinn *et al*, (1994) examined the sedimentary record of planktonic and sea ice diatoms (these organisms are important primary producers in the Antarctic and, as their siliceous cell walls are well preserved can often be identified to the species level). McMinn *et al.* (1994) found that the relative abundance of the various species had not changed significantly over the last 20 years, and certainly showed no discernible effect of the increasing UVB flux with time. This observation, has subsequently been challenged as being unrepresentative of the response of Antarctic diatom populations to elevated UVB (Bothwell *et al.,* 1995).

Any change in the species composition, particularly at the lower end of the foodweb, would have an impact at higher trophic levels (including pelagic fish) in terms of nutrient availability, recruitment and predator/prey relationships. It could also affect global biogeochemical cycles and hence the fate of anthropogenic increases in atmospheric CO_2.

Primary Producers: benthic

For benthic primary producers, there are data on the influence of UVB on photosynthesis in marine macrophytes (large phototrophs) Larkum and Wood (1993) found that deep-water (subtidal) Australian macroalgae and seagrasses were as sensitive to UVB as phytoplankton cells, and more sensitive than the shallow-water (intertidal) macroalgae tested. The macrophytes with their longer optical pathlength through photosynthetic organs have greater scope for screening UV out with intracellular UVB absorbing compounds, than phytoplankton cells (see Raven, 1991; Garcia-Pichel, 1994). It is not clear if the greater UVB sensitivity of the deep-water algae (Larkum & Wood, 1993) which are not normally exposed to high UVB fluxes is a result of their not capitalising on this potential for screening. Deep water populations of a given algal species have been found to be more sensitive to UVB than corresponding rock pool populations from European waters (Häder *et al.*, 1996).

However, benthic primary producers need not be large or marine; Bothwell *et al*. (1994) examined freshwater populations of benthic diatoms and showed that the initial UVB inhibition of growth rate, and hence biomass, of the diatoms was followed by an increase in their biomass as the impact of UVB on grazers (chironumid larvae) exceeded that on the primary producers.

Freshwater Fish

Studies in freshwater environments have examined the consequences of elevated UVB exposure for fish aquaculture (Bullock, 1988) and, in experimental studies on salmonids, for fish living in clear water alpine lakes (Little & Fabacher, 1994). Riverine fish and those living in clear alpine lakes are likely to be more vulnerable to UVB exposure and damage than wild marine species because clear fresh water lacks absorbing salts resulting in greater UVB penetration. By contrast, the presence of diverse humic (plant derived) substances affects penetration of UVB (10% of surface irradiance) in some freshwater environments, penetration may be reduced to as little as a few centimetres (Kirk, 1994b). Whilst attenuation of UVB may be greater in freshwater lakes, farmed fish are generally kept in comparatively shallow water at high stocking densities and so cannot swim deeper in order to avoid the solar radiation (Bullock, 1982). The absence of toxic contaminants in the freshwater environment of farmed fish may be something of a mixed blessing. On the positive side, farmed fish should not be exposed to photosensitizing contaminants which can enhance UVB potency whilst, on the negative side, the absence of such contaminants decreases the attenuation of UVB. There is, also, the possibility of enhanced UVB penetration in freshwater bodies resulting from acid rain induced reductions in DOC.

Intertidal benthic animals and fish

It is well established that the greatest impact of UVB is either at or immediately below the sea surface where eggs and early life stages of many commercially important species animals reside for some period of their lives. In addition young stages of many flatfish (e.g. 0 group turbot) live for part of their lives in the shallow waters (5-10 cms) of sandy tidal flats. As such they too would be exposed to virtually unattenuated UVB radiation. Intertidal benthic animals living on mud banks and beaches may need to bury themselves or be subjected to the full strength of UVB radiation. For nearshore animals, Kramer (1990) concluded that the relatively high concentrations of SPM and Gelbstoff would limit the transmission of UVB to less than 1% of the euphotic zone (<0.14 m) and, no biological effect would be expected. Whilst this expectation may be true for uncontaminated, highly turbid waters it does not appear to take into account the interactions discussed above between UVB and contaminants in, and immediately below, the surface microlayer, nor the observation by Döhler (1984), working with diatoms, that UVB can amplify the effects of other, natural, stressing factors such as salinity and temperature changes. Whilst some elements of physiological adaptation are possible in the biota that are subject to elevated UVB for protracted periods of time. Physiological adaptation is unlikely to occur in UK coastal waters because of the transient nature of intense UVB radiation (due to cloud cover), and the comparatively short residence time for some components of the biota in the upper levels of the water column.

4.3 Conclusions

UVB has its greatest impact either at or immediately below the surface of the water where primary producers as well as the eggs and early life stages of many commercially important species of animals reside for some period of their lives. In addition, the tidal and intertidal zones of marine and brackish environments, shallow streams and some fish farms are vulnerable to severe impact owing to their limited capacity to attenuate UVB radiation.

Whilst there is ample evidence to support the fact that exposure to elevated levels of UVB is detrimental to marine animals and plants there is very limited long term data from which to construct predictive models of effect for use in risk assessments. Furthermore, the consequences for marine life of the interactions between UVB and other anthropogenic factors remains unresolved; a situation further exacerbated by the limited understanding of the mechanisms of UVB damage for aquatic life. A knowledge of both the mechanisms of UVB damage and the interactions between UVB and environmental variables is central to the construction of reliable predictive models of effect for use in risk assessment and the formulation of research funding strategies and future policy.

In order to construct reliable predictive mathematical models for use in risk assessments the extent and mechanisms of UV effects on molecular and cellular processes, primary production and biodiversity are required. Biological understanding must be supported by information on the interactions between UV radiation and diverse natural environmental factors including water salinity, temperature and type (in terms of its UVB transparency) and man made inputs (including acid rain and organic contaminants). Finally, whilst the levels of UVA reaching the earth are not affected by ozone depletion its role in various biological processes remains unclear and should be addressed since it may be a confounding factor to any assessment of elevated UVB effects.

Damage attributed to UVB exposure for aquatic animals includes:

Sunburn*

Fungal infection

Altered pigmentation*

Decreased fecundity

Decreased growth and development

Decreased survival rate

Decreased mobility, impaired swimming activity

Inhibition of photoactic and photophobic responses

Changes in species composition

Skin* and gill pathology in fish

Restricted jaw movement

Irregularity in heartbeat

Dermal cell necrosis

Brain and retinal lesions*

DNA damage*

Reduction in ATP content

Membrane damage via free radical generation.

* also observed in man and animals

(modified from Kramer 1990)

5. Impacts On Terrestrial Ecosystems

- **UVB radiation affects many organisms which occur in terrestrial ecosystems. The magnitude of such effects in response to ozone depletion is poorly known except for some crop species.**
- **Increased UVB radiation can modify plant growth and development, leading in some crops to reduced yield. There is scope for plant breeding to develop crops with increased UVB tolerance, particularly in annual species, but this would have limited practical value for perennial species and natural vegetation.**
- **Mechanisms by which increased UVB is likely to affect terrestrial ecosystems are predominantly by altering competition between plant species and by interactions of plants with pests, pathogens and decomposers.**
- **The cumulative effects on terrestrial ecosystems of small but long-term increases in UVB are essentially unknown but may be significant for the function of native ecosystems and biodiversity.**

5.1 Introduction

The functioning of terrestrial ecosystems depends on a great diversity of organisms both above the ground and within the soil. However, until very recently research on terrestrial impacts of UVB has focused on plants and particularly on agricultural crops. The possible effects of increased UVB on higher animals are considered in Chapter 7 and in this Chapter we concentrate on the effects of UVB on plants. The focus of past research on green plants has some justification as plants are primary producers: by using the visible wavelengths of sunlight (400-700nm) to convert carbon dioxide into sugars (the process of photosynthesis), they provide the energy required by animals and which drives all other ecosystem processes. Photosynthesis requires sunlight, so that most plants would be subject to any increases in UVB. In intensive agriculture, the crop plants interact with pests, pathogens and soil organisms. In natural and semi-natural communities, such as forests, interactions may be more complex and include competition, herbivory, disease, predation and nutrient cycling (Figure 5.1). Any change resulting from increased UVB has the potential to affect not only agricultural food production and the functioning of natural ecosystems but also the biodiversity of organisms and our own perception of landscape and the natural world.

5.2 System responses and estimated impacts

Ecological research into the effects of UVB is a relatively new discipline. Until very recently studies have concentrated on agricultural crops which constrains any assessment of the impacts of increasing UVB on whole ecosystems. The literature concerning plant responses to UVB is substantial (see reviews by Caldwell *et al.*, 1989; Teramura *et al.*, 1991; Bornman & Teramura, 1993; Caldwell *et al.*, 1995) and it is clear that increased UVB may affect plants in many ways (see Chapter 3 for responses at the molecular and biochemical level). However, how these potential effects relate to ozone depletion is difficult to assess due to the wide range of methods that have been used. For example, studies in glasshouses or in artificially illuminated growth rooms may exaggerate plant responses to UVB and even field experiments may be hard to interpret (see Box 5; McLeod, 1996).

Box 5

Contrasting Methods For UVB Research

UVB effects on terrestrial ecosystems have been investigated using several techniques:

Controlled Environment and Glasshouse Studies

Experiments in plant growth rooms and glasshouses provide low levels of UVA (315-400nm) and visible radiation used in photosynthesis (photosynthetically-active radiation, PAR:400-700nm) which tends to increase the damaging effects of UVB (see Bornman & Teramura, 1993, Caldwell *et al.*, 1995). Such studies may also overestimate UVB effects by comparing doses substantially above those occurring in the field with an unrealistic "control" treatment of zero UVB.

Field studies

Studies in the field attempt to minimise changes to environmental conditions and have either reduced ambient UVB using filtration or supplemented ambient UVB using artificial lamps.

Field Filtration Studies

The simplest field filtration methods have used plastic screens to reduce ambient UVB levels. Effects observed beneath UVB transmitting filters have been compared to effects beneath filters opaque to UVB but such extreme comparisons may overestimate UVB responses. A more complex filtration method has utilized a layer of ozonized air above an outdoor cabinet.

Field Supplementation Studies

A large number of recent studies have used artificial UVB lamps to increase doses in the field. The lamps have frequently been operated by switching them on/off at a constant level and for a fixed period each day, the so-called square-wave" treatment, but this fails to maintain the normal relationship between UVB and PAR. Another problem with 'square-wave" treatments is that they rely on modelled estimates of UV to calculate the dose and equivalent ozone depletion. The models are complex (see Austin, Chapter 2) and depend on a number of assumptions, not least the validity of a specific action spectrum (see Chapter 1). Since the modelled estimates of UVB rarely take account of the effects of cloud, square-wave treatments provide substantially greater percentage additions than intended (Fiscus & Booker, 1995; McLeod, 1996). Modulated field supplementation studies have avoided some of these problems because they constantly adjust the UVB output of the lamps in relation to measured ambient levels of UVB, and are the preferred method of study. However, modulated systems are expensive and technically complex, and have only recently been deployed at a few sites in the UK and worldwide (McLeod, 1996).

Plant growth and development

UVB generally reduces plant growth (irradiated plants tend to be shorter, have smaller leaves and fewer branches) and modifies biochemistry (especially with respect to changes in protective pigments). However, there is marked variation in UVB responses, both between species and genotypes within species. There are many cases where increased UVB has no effect on plant growth and some where growth is significantly increased (see Caldwell *et al.*, 1995). Despite the detailed knowledge of the effects of UVB on many plant processes, the mechanism(s) underlying UVB responses in the field, whether inhibition or stimulation are poorly understood, but do not depend on reduced photosynthesis. UVB may also influence plant development. Flowering in irradiated plants may be reduced and/or delayed, and the number and/or weight of fruit may be reduced, by increased UVB.

Crop yield.

The effects of UVB on the commercial yield of crop plants, including those grown for animal feed, under field conditions are known from only a few species (Table 5.1). The bulk of the available data comes from older studies, which have not been repeated, and which often use UVB treatments which assume large ozone depletions (see review by Krupa and Kickert, 1989). Interpretation is also limited by the diversity of crop varieties and growing conditions used. Thus, although in many species there is a tendency for yield to be reduced under elevated UVB, the data are insufficient to draw general conclusions concerning changes under future UVB environments. Comprehensive data are confined to a few grain crops, either cereals or pulses (Table 5.1). In the context of UK agriculture, work on the responses of arable crops to UVB in the field started only in 1994, and remains confined to a single site (Mepsted *et al.*, 1996). The best studied crops, soybean and rice, have highlighted the magnitude of between-variety variation in response. In both species, yield is significantly reduced in many varieties but increased in a number of genotypes (see Caldwell *et al.*, 1995). Yield responses also vary from year to year, depending on weather conditions. For example, even in a sensitive soybean variety, yield was unaffected by UVB in years of low rainfall (Sullivan & Teramura, 1990).

Table 5.1 Effects of UVB on crop yields in the field.

Data are based on the reviews of Krupa and Kickert (1989) and Caldwell *et al.* (1994) and more recent papers listed below. UK field crops are shown in bold.

Crop	Yield		
	Decreased	**No change**	**Increased**
1. Data from more than one study including modern exposure methods			
Pea[1]	✓	✓	
Rice	✓	✓	✓
Soybean[2,3]	✓	✓	✓
2. Data from more than one study			
Beans (*Phaseolus* spp.)		✓	✓
Maize	✓	✓	✓
Peanut	✓	✓	
Potato	✓	✓	
Radish		✓	
Tomato	✓	✓	✓
Wheat	✓	✓	
3. Data from single studies			
Barley[4]		✓	
Millet		✓	
Sorghum		✓	
Sweet corn	✓		
Cowpea	✓		
Cotton		✓	
Pepper	✓	✓	
Squash	✓		
Broccoli	✓		
Mustard	✓		
Tobacco		✓	
Cassava[5]	✓		
Turnip	✓		

[1] Mepsted *et al.* (1996)
[2] D'Surney S.J. *et al.* (1993)
[3] Fiscus, E.L, Miller, J.E., & Booker, F.L. (1992)
[4] Corlett *et al.* (1997)
[5] Ziska *et al.* (1993)

Figure 5.1

Components of terrestrial ecosystems that may respond to UVB.

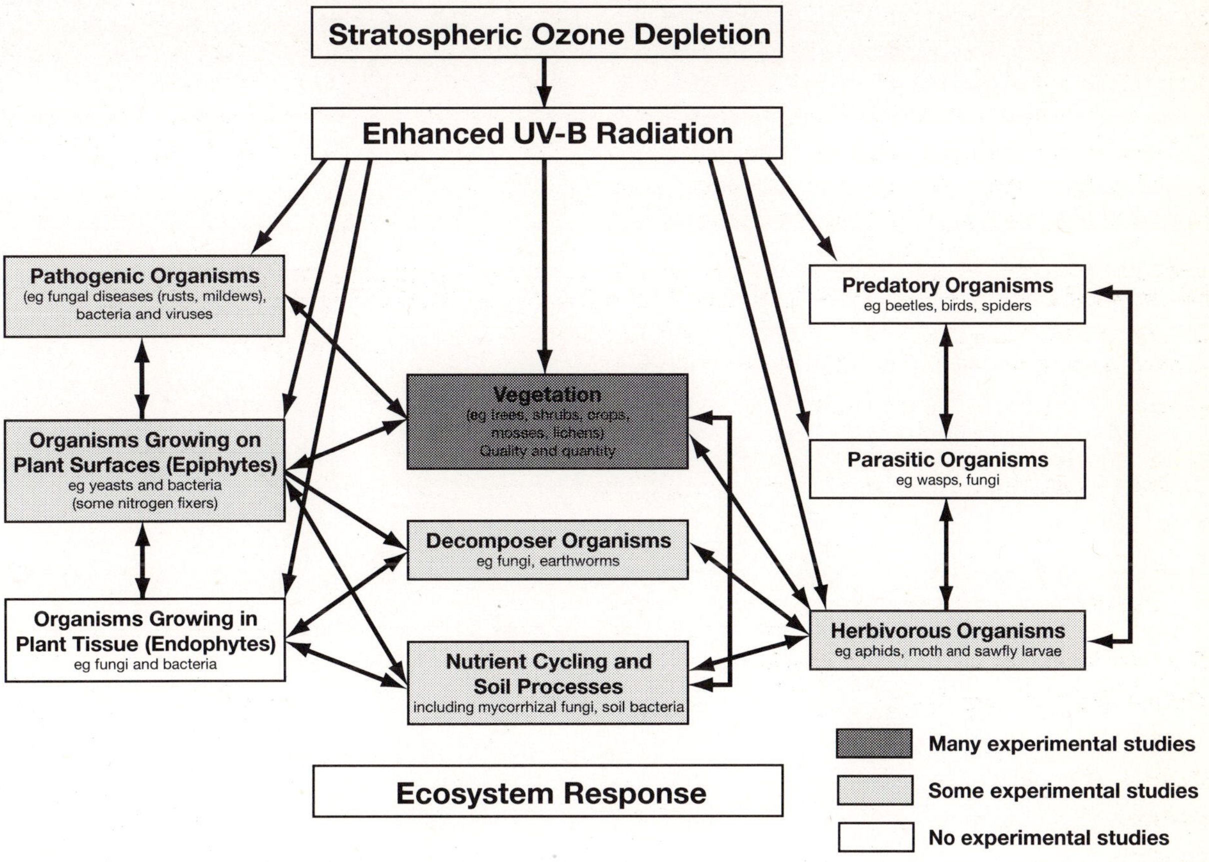

Interactions of plant responses with other environmental factors and other organisms

Although data remain relatively limited, plant responses to UVB may be modified by other factors. Mild drought, nutrient deficiency and elevated temperatures may reduce growth reductions caused by UVB (see Bornman & Teramura, 1993, Caldwell *et al*., 1995). In wheat and rice, but not soybean or *Pinus taeda*, UVB inhibited the normal increase in photosynthesis and growth under elevated carbon dioxide concentration (Rozema *et al*., 1996) There are also some observations that the effects of UVB on plants may interact with those of increased ozone concentrations in the lower atmosphere (Rao & Ormrod, 1995).

In the field, relatively small effects of UVB on individual plants may be exaggerated by competition between plants. In crops, competitive balance may change in favour of either the crop or the weed (Barnes *et al*., 1995). Alteration of the competitive balance between native plants may influence the complex interactions occurring in species-rich natural vegetation, but has rarely been studied (e.g. Gold & Caldwell, 1983).

Any change in interactions between plants and pathogens or pests due to increased UVB will be a function of the effects of UVB on each organism involved. Thus, UVB may affect decomposition and nutrient cycling in an ecosystem both through changes in properties of plant leaf litter and by direct effects on decomposer fauna and microbes (Figure 5.1). Fungi play a key role in decomposition, and cause many important plant diseases. Significant responses to both UVB and also UVA radiation have been observed in many fungi, but most studies have not been concerned with ozone depletion and cannot be easily related to the field. Recent, more realistic studies have confirmed that UVB responses may be inhibitory or stimulatory (Paul *et al*., 1996). The direct effects of UVB on pathogenic fungi, combined with increases in antimicrobial plant chemicals are often expected to reduce disease. However, while such decreases have been observed, increased disease under elevated UVB have been reported more frequently (Manning & Teidemann, 1995; Paul *et al.*, 1996). A recent study has also demonstrated that UVB increases can modify the below-ground communities, including mycorrhizal fungi, associated with the roots of trees (Klironomos & Allen, 1995). In the only study to have considered the effects of UVB on decomposition (Gerhke *et al.*, 1995), leaf litter of arctic shrubs grown under elevated UVB in the field decomposed more slowly, and increased UVB during decomposition reduced the colonisation of litter by decomposer fungi. Direct effects of environmental UVB on insects are not known but it may damage pathogenic organisms which cause diseases of insects (Paul *et al.*, 1996). However, increased UVB can induce changes in host plant chemistry to affect the quality of animal feed and herbivory by certain insects. Although such mechanisms have been observed, they have not been evaluated in the context of changes in UVB dose resulting from ozone depletion.

Other terrestrial organisms

The responses of most terrestrial vertebrates to increasing UVB remains almost completely unknown (but see Chapter 7 for domesticated livestock). The resistance of amphibian eggs to UVB radiation damage has been investigated in North America using the field filtration of incident sunlight (see Box 5) over small enclosures in natural habitats. The hatching success of embryos was significantly increased by filters which removed ambient UVB (Blaustein *et al.*, 1995). The effects of UVB have also been shown to interact with pH and fungal infection of the eggs (Long, Saylor & Soule, 1995; Kieseker & Blaustein, 1995). These observations support the hypothesis that increased UVB radiation contributes to a decline in some amphibian populations.

Sensitivity of system/sector

It has been argued that crop photosynthesis is not sensitive to realistic UVB treatments in the field (Fiscus & Booker, 1995). However, crop productivity is not determined solely by photosynthetic rates, and until the effects of UVB on plant morphology

Box 6

Problems associated with plant action spectra

Most of the published literature relating to plant responses to UVB has been based upon a generalised plant action spectrum (PAS) devised by Professor Martyn Caldwell (Caldwell, 1971). This action spectrum was based upon a consideration of a range of nine published responses but, notably, data on long-term plant growth was not available. The generalised plant action spectrum was initially used in the form of a mathematical function fitted by Green *et al.* (1974), which decreases to zero at 313nm, and which is normalised to 1 at 280 nm (PAS280). PAS280 was used in many early publications including the 1989 UNEP report (Caldwell *et al.*, 1989). However, subsequently the generalised plant action spectrum was expressed normalised to 300nm (Caldwell *et al.*, 1986), and this is the form in general use today (PAS300). However, care must be exercised when interpreting older literature, as the normalisation wavelength of the generalised plant action spectrum may not be readily apparent.

Some confusion has occurred because a second mathematical function has been fitted to Caldwell's original data set, and this spectrum (Thimijan *et al.*, 1978) and has been used as PAS300 in Europe. This function rises more steadily with decreasing wavelengths in the UVA than the original PAS300, and has a long "tail" in the UVA waveband. As noted in Chapter 1, such a difference in action spectra produce different estimates of absolute PAS300 doses and, more importantly, different radiation amplification factors (see Chapter 1) particularly in the winter months (McLeod, 1996). Thus, different publications, all specifying "PAS300" may actually be using substantially different treatments. Since other plant action spectra have been published, generally referring to particular processes such as pigment formation of DNA damage, further scope for uncertainty in the design and interpretation of experiments clearly exists.

and consequently light interception, or the prevalence or severity of pests and diseases are known, many crops must still be considered as potentially susceptible to increased UVB. Even for the best studied mechanisms, the relationship between UVB dose and crop responses are too poorly defined to quantify the magnitude of any impacts in relation to the ozone depletion anticipated in the UK.

The potential sensitivity of natural and semi-natural communities for significant UVB impacts, is probably greater than that of crops. Since many characteristics of natural systems are functions of long-term processes acting over many decades, it is possible that quite subtle changes resulting from increased UVB may accumulate over time to produce significant effects. However, the absence of pertinent dose response data precludes any quantitative modelling and assessment of the magnitude of such impacts.

5.3 Potential adaptation where known

The variation in response between varieties of many crop species provides scope for deliberate selection for increased UVB tolerance. In many annual crops, some varieties currently in commercial use are tolerant to elevated UVB levels. Increased use of such varieties provides a short-term option, should moderate ozone depletion threaten yields over the next few decades. However, data for UK varieties remain limited and the scope of this approach under UK conditions is not clear. It may therefore be advisable to assess varieties currently under development for their response to potential future UVB climates. Deliberate breeding of tolerant varieties may be necessary should more severe ozone depletions persist. However, a practical basis of selection remains unclear given the uncertainty concerning key mechanisms of response in the field (see above), and screening in growth rooms or using specific biochemical or molecular traits may prove unreliable. Direct screening of UVB responses using field techniques would be preferable but is not without technical difficulties (Boxes 5 and 6).

For perennial crops, including many commercial fruits as well as forestry trees, intervention to moderate UVB effects may not be feasible. The vulnerability of these crops to elevated UVB is not currently known and if future research indicates cause for concern, any deliberate breeding programme would probably be too slow to moderate the effects of anticipated ozone loss. Selection for increased UVB tolerance in such crops may prove to be necessary if the extent and duration of ozone loss exceeds current estimates.

Increasing UVB tolerance by selecting among existing varieties, or breeding new varieties, may be applicable to some crops but is not appropriate to natural vegetation. Even if UVB does not affect the overall productivity of natural ecosystems, changes in individual species or groups, even if small in magnitude or of short duration, may be amplified by competitive interactions to produce long-term changes in biodiversity. Elevated UVB will increase the selective pressure on populations in favour of individuals with physiological, structural and, in animals, behavioural characteristics which increase tolerance or avoidance of UVB. However, the long life cycles of some organisms (e.g. forest trees) may preclude rapid selection of tolerant individuals. The rapidity of changes in UVB may exceed the potential of organisms to adapt before numbers of particular species are affected. Processes with long time scales, such as biogeochemical cycles, which may be vulnerable to UVB effects on decomposition and nutrient cycling, may also have a detrimental impact on some natural ecosystems with little opportunity for selection and adaptation.

5.4 Uncertainties and unknowns

The design and interpretation of experiments, as well as expected changes in the biologically weighted UVB climate, depends on the action spectrum used (see Chapter 1). For plants, markedly different action spectra have been determined for different processes, although most studies dealing with whole plant responses have adopted the "Generalised plant action spectrum" (see Box 6). However, uncertainty over the validity of this action spectrum and the use of different mathematical formula to describe this spectrum (see Box 6), is a fundamental uncertainty in the interpretation of existing data and future studies.

The understanding of biological responses to UVB still relies substantially on short-term studies from glasshouses and growth rooms. While such studies can provide valuable insight into mechanisms of UVB response, their value is often limited by the use of unrealistic growing conditions and sometimes by extreme (and inadequately defined)

UVB doses. UVB dose responses remain virtually uncharacterised, both qualitatively and quantitatively, even in well-studied plant systems.

Understanding the effects of increasing UVB on terrestrial ecosystems requires well-designed field studies. Unfortunately, uncertainties exist in integrating data from different forms of field experiment (see Box 5 and McLeod, 1996). The interpretation of field studies is also difficult as few extend beyond a single season or a single generation and consequently, we have no understanding of the potential long-term effects of realistic increases in UVB. This is especially significant for understanding non-crop systems where responses may be slow and cumulative and where the lack of data on effects on organisms other than higher plants amplifies the uncertainty over the responses of native ecosystems

5.5 Principal implications for other sectors and the UK economy

Implications for agricultural production

In the light of existing evidence, large reductions in yield of arable crops are unlikely given current predictions for ozone loss for the UK. However, significant effects on arable crops cannot yet be excluded: well-founded field data is still too limited to refute the numerous reports that crop growth and yield can be reduced by increased UVB. Significant intra-specific variation in UVB sensitivity may prevent the use of particular varieties under future UVB climates. Implications for production of perennial crops, including forestry trees, are less clear. However, significant reductions in the growth of several conifers under experimental UVB treatments, suggests that significant effects cannot be excluded.

Implications for biodiversity and landscape

Worldwide, there is little published data dealing with the effects of UVB on native plants and natural or semi-natural ecosystems: in the UK there are none. Therefore, any assessment of the implications of ozone depletion on biodiversity or landscape in the UK must be treated as preliminary. In the light of current information and general ecological principles, significant impacts cannot be excluded. However, at present, the magnitude and nature of any effects cannot be defined.

5.6 Conclusions

Some of the unresolved aspects of UVB responses (identified in Section 5.4) are being addressed under current UK research projects. Future research can build on these foundations but the following significant areas of uncertainty remain:

- Knowledge of plant responses to realistic UVB doses under field conditions and their underlying mechanisms is limited and must be characterised for a range of ozone depletions scenarios using the best available methodologies.
- The responses of populations, communities and ecosystems to elevated UVB are poorly known and no assessment of possible effects on biodiversity is possible. Future research must consider the extent and mechanisms of UVB impacts on interactions between organisms and on processes other than primary production by higher plants.
- Given the technical complications of UVB exposure methods (see Boxes 1, 2, 5 and 6) future research requires a proper understanding of appropriate action spectra for long-term responses of a wide range of organisms and processes. The lack of information on relevant action spectra indicates that assessment of alternative methodologies for UVB investigations would be worthwhile.
- Scaling-up future results from UVB impacts studies to a national scale is limited by a lack of information on existing UVB climates in the UK, e.g. temporal variation (due to clouds, season and natural fluctuations in ozone) and by geographical variation.

There is a wide and diverse body of research on the effects of increased UVB on terrestrial ecosystems, but there are many gaps which remain in existing understanding. Although, recent research developments will improve understanding, it is regrettable that few definite conclusions can be derived from the existing literature. Current research does not support earlier predictions that ozone depletion will result in acute UVB injury to plants and large-scale reductions in crop yields. Rather, attention is now focused on the more subtle effects of UVB. These include changes in plant morphology, the chemical composition of plants, interactions between plant species and other organisms (e.g. pests and pathogens) and decomposition. Although not dramatic in the short-term, such effects have the potential to cause long-term impacts to biodiversity and biogeochemical cycling, particularly in natural ecosystems. The possibility of significant impacts of UVB on terrestrial ecosystems cannot be excluded.

6. Human Health Impacts

- **In principle, a reduction in ozone levels could increase the incidence of harmful effects on health that are known to be caused by the UVB radiation in normal sunlight. The principal deleterious effects of UVB radiation are:**
- **Skin damage, including sunburn, photoageing and skin cancer**
- **Eye abnormalities, including snowblindness and cataract.**
- **Changes in the immune system, which may alter the course of cancer or infectious disease.**

UVB can damage the skin and eyes, and cause changes in the immune system.

UVB-induced immunosupression may be a factor in human skin cancer and may modify infectious disease processes.

Skin cancer incidence in the UK is rising rapidly but there is no evidence that ozone depletion has contributed to this.

Ozone depletion expected over the next few decades is predicted to contribute to an additional lifetime risk of skin cancer in children alive today of between 4 to 10% greater than expected in the absence of ozone depletion.

6.1 Effects on Skin

Sunburn

Erythema, ("sunburn") or redness of the skin due to dilatation of superficial dermal blood vessels, is one of the commonest and most obvious effects of ultraviolet exposure. Following exposure to solar UV radiation, there is usually a latent period of 2 to 4 hours before erythema develops. Erythema reaches maximum intensity between 8 and 24 hours after exposure, but may take several days to resolve completely. If a high enough exposure has occurred, the skin will also become painful and oedematous, and blistering may result.

Tanning

Another consequence of exposure to solar UV (which at present still seems to be socially desirable) is the delayed pigmentation of the skin known as tanning, or melanin pigmentation. Melanin pigmentation of skin is of two types: (i) constitutive - the colour of the skin seen in different races and determined by genetic factors only; and (ii) facultative - the reversible increase in tanning in response to sun exposure.

Hyperplasia

In addition to tanning, the skin is capable of another perhaps even more important adaptive response which limits damage from further ultraviolet exposure - epidermal thickening or hyperplasia. This begins to occur around 72 hours after exposure. It is a result of an increased rate of cell division in the lower epidermis, and eventually results in thickening of both epidermis and stratum corneum (the outermost layer of the epidermis) which persist for several weeks. This adaptive process, unlike tanning, does not depend on a genetic predispostion and is the major factor which protects those who tan poorly in sunlight.

Production of vitamin D

The only well-established beneficial effect of solar UV on the skin is the production of vitamin D_3. The skin absorbs UVB radiation in sunlight to convert sterol precursors in the skin, such as 7-dehydrocholesterol, to pre-vitamin D_3 which is then transformed into Vitamin D_3. Vitamin D_3 is further modified by the liver and kidneys to biologically active metabolites such as 25-hydroxyvitamin D; these metabolites then act on the intestinal mucosa to facilitate calcium absorption, and on bone to facilitate calcium exchange.

Skin cancer

The three common forms of skin cancer, listed in order of increasing seriousness are: basal cell carcinoma (BCC), squamous cell carcinoma (SCC) and malignant melanoma (MM). Around 90% of skin cancer cases are of the non-melanoma variety (BCC and SCC) with BCCs being approximately four times as common as SCCs. Exposure to UV radiation is considered to be a major etiological factor for all three forms of cancer (Armstrong and Kricker 1995; IARC monographs, 1992). For basal cell carcinoma and malignant melanoma, neither the wavelengths involved nor the exposure pattern that results in risk have been established with certainty; whereas for squamous cell carcinoma, both UVB and UVA are implicated and the major risk factors seem to be cumulative lifetime exposure to UV and a poor tanning response.

Squamous cell cancer

The evidence that exposure to sunlight, even without ozone depletion, is the predominant cause of squamous cell cancer in man is very convincing. These cancers occur mainly on sun-exposed skin such as the face, neck and arms, and the incidence is clearly correlated with geographical latitude, being higher in the more sunny areas of the world (Kricker *et al.,* 1994). Recent epidemiological studies suggest that sun exposure in the 10 years prior to diagnosis may be important in accounting for individual risk of SCC (Gallagher *et al.,* 1995a).

Basal cell cancer

The relationship between basal cell carcinoma and sunlight is less compelling, but the evidence is sufficiently strong to consider it also to be a consequence of exposure to sunlight. Whilst SCC is strongly related to cumulative lifetime exposure to sunlight, this relationship is not so convincing for BCC (Gallagher *et al.,* 1995b; Kricker *et al* 1995), and it may be that sun exposure in childhood and adolescence may be critical periods for establishing adult risk for BCC (Gallagher *et al.,* 1995b).

Malignant Melanoma

During the past 40 years or so there has been an increase in the incidence of malignant melanoma in white populations in several countries. There exists an inverse relationship between latitude and melanoma incidence and this has been taken as evidence for a possible role of sunlight as a cause of malignant melanoma. However, this pattern is not always consistent. In Europe, for example, the incidence and the mortality rates in Scandinavia are considerably higher than those in Mediterranean countries. This inconsistency may reflect ethnic differences in constitutional factors and customs. Also, the unexpectedly low incidence in outdoor workers, the sex and age distribution, and the anatomical distribution have pointed to a more complex association (Armstrong and Kricker 1994).

There is now growing evidence that intermittent sun exposure - mainly from recreational activities - rather than cumulative or chronic exposure associated with occupation is associated with increased risk of developing malignant melanoma. Several studies have established a history of sunburn as an important risk factor for melanoma development, although in these studies a potential for recall bias exists. Migration studies have led to the suggestion that sun exposure in childhood is a particularly critical period in terms of melanoma risk (IARC, 1992).

Risk analysis of human skin cancer

Estimates of the risk of inducing skin cancer from exposure to ultraviolet radiation demands knowledge of dose-response relationships and the relative effectiveness of different wavelengths of sunlight in causing skin cancer. Data on dose-response relationships and action spectra are available to some extent to allow quantitative estimates of the effects of ozone depletion on non-melanoma skin cancer (NMSC) incidence. These data remain unknown for malignant melanoma and so it is unwise to make predictions about the consequences of ozone depletion on this disease. It cannot be assumed, however, that because knowledge does not allow risk estimates to be made for malignant melanoma, ozone depletion will have no effect on the incidence of this disease.

Dose-response relationships

Application of multivariate analysis to population-based epidemiology of NMSC has shown that, for a group of subjects with a given genetic susceptibility, age and environmental ultraviolet exposure are the two most important factors in determining the relative risk. Other epidemiological studies have confirmed these findings, and this has led to a simple power law relationship which expresses the risk in terms of these factors (see Box 7). In most instances changes in lifestyle with age mean that the annual UV exposure does not remain constant and so more complex mathematical models of NMSC risk have been developed (see Box 7).

Action spectrum for photocarcinogenesis

Clearly an action spectrum for skin cancer can only be obtained from animal experiments. The most extensive investigations to date are those from groups at Utrecht and Philadelphia. These workers exposed a total of about 1100 white hairless mice to 14 different broad-band ultraviolet sources and by a mathematical optimisation process derived an action spectrum referred to as the *Skin Cancer Utrecht-Philadelphia (SCUP)* action spectrum (de Gruijl *et al.,* 1993). The SCUP action spectrum is that for skin tumour induction in hairless mice, a species which has a thinner epidermis than humans.

By taking into account differences in the optics of human epidermis and hairless albino mouse epidermis, the experimentally-determined action spectrum for tumour induction in mouse skin can be modified to arrive at a postulated action spectrum for human skin cancer (de Gruijl and van der Leun 1994). The resulting action spectrum resembles the action spectrum for erythema (McKinlay and Diffey, 1987) (Fig. 6.1) suggesting that this action spectrum may be used as a surrogate for human skin cancer. For this reason erythemal doses expressed in units of *minimal erythema dose (MED)* can be used to express carcinogenic-effective exposure. It is important to appreciate, however, that the MED is used as an exemplary measure of *exposure* rather than as the dose necessary to cause a biological outcome in any given individual. Here 1 MED is defined as equivalent to an erythemally-weighted radiant exposure of 200 J/m^2 which has been shown to approximate the exposure dose necessary to result in a barely perceptible erythema in unacclimatized white skin.

Box 7

Simple mathematical model of non-melanoma skin cancer assuming steady state exposure throughout life

$$\textbf{risk} \propto (\textbf{annual UV dose})^{\beta} (\textbf{age})^{\alpha}$$

The symbols α and β are numerical constants associated with the age dependence of the incidence and the biological amplification factor, respectively, and are normally derived from surveys of skin cancer incidence and ultraviolet climatology.

More complex model taking into account changes in lifestyle with age

$$\textbf{risk} \propto (\textbf{cumulative UV dose at age T})^{\beta-1} \sum_{t=0}^{T} (\textbf{annual dose at age [T-t]}).t^{\alpha-\beta}$$

Figure 6.1

Action spectra for erythema (dashed line; McKinlay and Diffey 1987) and for non-melanoma skin cancer (solid line; de Gruijl and van der Leun 1994) in human skin

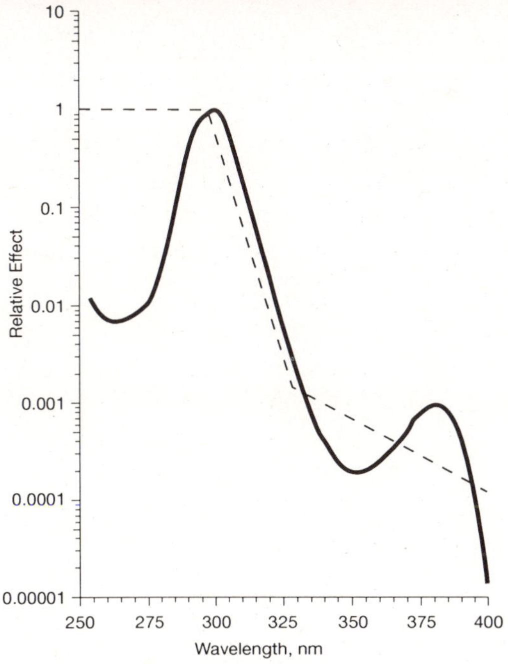

Human exposure to natural UV radiation

The solar UV radiation exposure received by an individual depends on three factors:

- ambient solar ultraviolet radiation
- the fraction of ambient exposure received on appropriate anatomical sites
- behaviour outdoors

Estimates of personal exposure can be obtained in two ways: by direct measurement using UV sensitive film badges normally worn on the lapel site, or by modelling the variables above. The results obtained from both methods indicate that indoor workers in the United Kingdom receive around 100 MED per year mainly from weekend and vacational exposure, and principally to the hands, forearms and face. This value is approximately 6% of the total ambient available. Children have a greater opportunity for outdoor exposure and receive an annual dose of around 150 MED. For indoor workers the annual exposure associated with occupation (travelling to and from work, going outside at lunchtime) is about 20 MED, about 40 MED is contributed by weekend exposure, and the remaining 30 MED from vacational exposure. In the case of children, 'occupational' exposure (playtime and lunchtime exposure) may be about 30 MED, recreational about 90 MED (because children are at school for only about 190 days per year), and vacation with parents giving about 30 MED. It must be stressed, however, that there will be large variations in the annual exposure doses received by individuals within a given population group depending upon propensity for outdoor activities and to what extent these are influenced by shade. This has been demonstrated in a recent large-scale study of the outdoor ultraviolet exposure of 180 children and adolescents in 3 geographically distinct regions of England over a period of three months (Diffey *et al.,* 1996).

Stratospheric ozone depletion and the risk of non-melanoma skin cancer in a British population

A 1% decrease in ozone will lead to a 1.2 to 1.4% increase in carcinogenic-effective UVB radiation [see Box 2 on RAF] (Health Council of the Netherlands 1994). For every 1% increase in carcinogenic-effective UVB radiation, it is estimated (Health Council of the Netherlands 1994) that there will be an approximate 2.5% increase of SCC incidence, with a corresponding figure of approximately 1.5% for BCC [see Box 2 on BAF]. Combining these we arrive at overall amplification factors, which can be summarised as:

$$1\% \downarrow \text{ in O3} \longrightarrow 1.4 \times 2.5 = 3.5\% \uparrow \text{ SCC}$$
$$\longrightarrow 1.4 \times 1.5 = 2.1\% \uparrow \text{ BCC}$$

We can use this approach to estimate the consequences of a sustained 10% stratospheric ozone depletion in the United Kingdom. The most recent figures for skin cancer incidence in the UK are based on 1989 data and are obtained from the Office of Population Censuses and Surveys for England and Wales, Information and Statistics Division of the NHS Directorate of Information Services for Scotland, and the Northern Ireland Cancer Registry. These sources yield a combined figure of just over 40,000 cases of skin cancer annually in the UK. Of these about 30,000 are BCC, about 6000 are SCC and the remaining 4000 are malignant melanoma. However, because of under-recording of non-melanoma skin cancers (BCC and SCC) in the UK and the rising incidence, the number of these cancers occurring each year is probably closer to double these figures. So, for a sustained

10% ozone depletion, we might expect an increase of 21% and 35% in the incidence of BCC and SCC, respectively, giving an additional 12,000 cases of BCC and 4000 cases of SCC each year in the UK.

Whilst this approach may be applicable to future generations, it is not appropriate to British people alive today. By combining ultraviolet climatological data for the United Kingdom with models of human behaviour in order to estimate annual exposure throughout life, and then employing the mathematical models described above, it is possible to estimate the lifetime risk of NMSC. Perturbation of climatological UVB expected as a consequence of ozone depletion presently occurring is also incorporated to examine the effects of environmental change on skin cancer incidence (Diffey, 1992).

For British adults alive today ozone depletion continuing indefinitely at current rates is predicted to result in a relatively small (<5%) additional lifetime risk of non-melanoma skin cancer, assuming no changes in climate, time spent outdoors, behaviour or clothing habits (NRPB, 1995). The lifetime risk incurred by today's children, however, is predicted to be 10-16% greater than expected in the absence of ozone depletion. However if the production and use of substances which deplete ozone are reduced as expected under the current provisions of the Montreal Protocol the increased lifetime risk of skin cancer is likely to be less than these estimates, typically between 4 to 10%. It is to be hoped that public awareness about the adverse health effects of sun exposure will achieve these changes, which could lead to a reduction - rather than the anticipated increase - in skin cancer incidence. If there is no change in sun exposure habits and the rate of ozone depletion the calculations suggest that 50 years from now the number of skin cancers occurring each year in the UK will increase from the present number of about 80,000 to around 88,000. This is a much smaller increase than has occurred over the previous 50 years and illustrates that changes in leisure time, fashion and activities in the sun - factors which are commonly believed to be important in the rising incidence of skin cancers - have had a much greater effect on skin cancer rates than expected as a consequence of ozone depletion. Clearly it will prove difficult to identify the real effect of ozone depletion on skin cancer incidence over the next few decades.

Photoageing

Solar exposure causes skin photoageing which is different from normal chronological ageing. Clinically, this is characterised by wrinkling, loss of elasticity and localised alterations of pigmentation on habitually sun-exposed sites. The most evident histological marker is elastosis, characterised by dermal elastic tissue hyperplasia and loss of fine structure. The most thorough action spectrum study for photoageing (elastosis) in mice shows a spectral dependence similar in general shape to that for non-melanoma skin cancer in mice (Kligman and Sayre, 1991). Thus, the increased risk of photoageing as a result of ozone layer depletion in the UK is likely to be similar to that for skin cancer.

6.2 Eye damage

Exposure to sunlight is believed to be associated with a variety of eye disorders, including damage to the cornea, lens and retina (Young, 1994). Cataracts are the most frequent effect and studies of their relationship with UVB exposure have been recently reviewed by Taylor (1995). Several epidemiological studies have shown a relationship between cataract incidence and ambient UVB levels but only two studies have shown a relationship with individual UVB exposure. This is consistent with experimental data.

The relative importance of different wavelengths in causing eye damage as well as dose relationships, are extremely uncertain. Also other factors such as diet play a role in the aetiology of cataracts and because of this we do not feel justified in making quantitative estimates of the consequences of ozone depletion on the incidence of ocular damage from sunlight exposure.

6.3 Immunosuppression

The skin is an important part of the immune system. It contains populations of cells, such as epidermal Langerhans cells (LC) and dermal macrophages, which process antigens encountered by skin and initiates an immune response in other internal immunological organs as well as skin. LC are very UV sensitive. UV exposure also results in the release of soluble inflammatory mediators (cytokines) which modulate immune function (Ullrich, 1995). Systemic UV-induced immunosuppression is usually demonstrated *in vivo* in humans and mice by the inhibition of an allergic reaction to an antigen (e.g.

chemical or pathogen). This inhibition is antigen specific. Subsequent exposure to the antigen will no longer induce an allergic response, even though responses to other antigens are unaffected. The sensitising process may be by topical application of small molecular weight substances called haptens (termed contact hypersensitivity (CHS)) response, or by subcutaneous injection of larger molecles or microbial infections (termed delayed-type hypersensitivity (DTH)).

Skin cancer

Evidence for the importance of overall immune function comes from renal transplant patients, maintained on immunosuppressive drugs to prevent rejection, who have a much higher incidence of non-melanoma skin cancer than normal (McGregor and Proby, 1995). UV-induced immunosuppression plays an important role in skin cancer in mice (Grabbe and Granstein, 1995). It is suspected of playing such a role in humans but the evidence is limited and indirect. Inhibition of CHS is regarded as a model of the photoimmunological events in skin cancer. Using this model, it has been reported that humans, irrespective of skin type, can be classified as UV-resistant and UV-sensitive with a distribution of approximately 1:1. However, more than 90% of skin cancer patients were UV-sensitive (Streilein 1995). Repeated sub-erythemal UV exposure has a cumulative immunosuppressive effect in humans (Cooper *et al.,* 1992) indicating that low dose UV exposure may have a significant effect.

Infectious diseases

The skin provides an entry route for many pathogens and it is also the organ in which many micro-organisms replicate. The effects of UV on the pathogenesis of infection have been studied because of the direct effects on the skin by UV exposure and the indirect effects on systemic immune function mediated via the skin. Although, studies have yet to address this issue, there is some concern that the efficacy of some microbial vaccines may be reduced by high UV.

There are few studies on humans but Armstrong (1994) recently concluded that most of current evidence for the harmful effects of solar UV on human immune function is inadequate. However, there is evidence that exposure to sunlight or artificial sources of UV radiation triggers the recrudescence of herpes simplex virus lesions (Klein and Linnemann, 1986; Spruance, 1985; Wheeler, 1975; Rooney *et al.*, 1991). In addition many SCC on UV-exposed sites in immunosuppressed individuals occur in areas with histological and virological indications of papillomavirus infection. Pathological and epidemiological studies in patients infected with *Mycobacterium leprae* indicate that increased exposure to sunlight leads to an increase in the incidence and severity of leprosy; most primary lesions occur on areas of the body not protected from sunlight.

Several animal models involving a range of viral, bacterial, fungal and parasitic pathogens have been developed in which the effects of UV radiation on pathogenesis can be examined (for reviews see Jeevan *et al.*, 1995; Norval *et al.*, 1994). The general approach of these studies has been to assess the hypersensitivity responses and other immunological parameters, together with the severity of the infection and the number of infectious micro-organisms produced. Examples of skin-associated infections have included the herpes simplex virus and the parasite *Leishmania major*, both in mice. Other models have concentrated on systemic non-skin associated infection such as cytomegalovirus, *Listeria monocytogenes* and *Trichinella spiralis* in rats and murine leukaemia virus, the yeast *Candida albicans* and *Mycobacterium bovis* in mice. In all cases UV-induced suppression of immune responses to the particular micro-organism was shown, although this did not always result in increased severity of clinical symptoms.

An attempt has been made recently to compare the susceptibility of the immune response of different species to UVB-induced changes (Goettsch *et al.*, 1995); basal immune function in rats, mice and humans were found to be modified in a similar way. Therefore it is probably valid to extrapolate animal data to humans for risk assessment as a consequence of ozone layer depletion when other parameters, such as wavelength dependence, are better defined.

Chromophores and Wavelength Dependence

There is experimental evidence to support urocanic acid (UCA), located in the stratum corneum (El-Ghorr and Norval, 1995), and epidermal DNA (Kripke *et al.,* 1992) as chromophores for the effects of UV on immune function. Both show maximal absorption in the UVC region with a tail in the solar UVB region and very little UVA absorption.

There have been no studies on the wavelength dependence effects on infectious agents. Action spectra have been determined for inhibition of CHS in mice, and antigen presentation in human epidermal cells *in vitro*. These show a maximal effect in the UVC region but immunosuppression is readily induced by UVB in the solar range (295 - 315nm) (De Fabo and Noonan, 1983; Elmets *et al.,* 1985; Hurks *et al.,* 1995). Using a mouse model, De Fabo *et al.* (1990) have calculated that at 60°N in winter, there would be a 1% increase in biologically effective dose for immunosuppression for every 1% decrease in ozone. Thus increased terrestrial UVB irradiance as a result of ozone depletion may modify immune function. However, the present state of knowledge does not allow us to reliably predict the effects of ozone depletion on the course of infectious agents or skin cancer in humans.

6.4 Other cancers

There is some evidence that solar radiation may play a role in ocular melanoma and cancer of the lip. However both cancers are rare and there is not the clear relationship with geographical latitude that is seen with skin cancers. Non-Hodgkin's lymphoma has increased in incidence in several countries over the past few decades with similar geographical and temporal trends to those seen in non-melanoma skin cancers raising the possibility of a link with sun exposure (Zheng *et al.,* 1992; Cartwright *et al.,* 1994). Further support for this hypothesis comes from observations that both melanoma and non-melanoma skin cancers occur appreciably more often than would be expected among patients who have previously been diagnosed as having non-Hodgkin's lymphoma (Adami *et al.,* 1995).

7. Impacts on Animals

- **Increased UVB may have direct effects on animal health, although most effects will be influenced by the ecological and social systems in which the animals live. Effects are likely to vary between farmed animals, companion animals and wildlife.**
- **Potential effects to animal health include an increased incidence of sunburn, skin cancer, degenerative eye conditions and changes to the immune system.**
- **Many of the more significant effects of increased UVB will be indirect and result from changes to the quality of nutrition and availability of shelter for farmed animals, which in turn affect productivity and management regimes. Adaptation to changing UV levels by farmed animals may be achieved through appropriate management; whilst wildlife may be more vulnerable and have to respond naturally.**
- **So far, the effect of increased UVB on animals is poorly understood, with the exception of some laboratory animal studies that have been undertaken to assess the potential effects of increased UVB levels on man. It is important to understand the effects of increasing levels of UVB on different animals and their responses under managed and natural environments.**

7.1 Introduction

Increased UVB may have direct effects on animal health in the UK, although the effects are likely to be influenced by the ecological and social systems in which the animals exist. Indirect effects may also result from changes within the food chain and altered seasonal and diurnal patterns which in turn affect animal activity. The indirect effects of increased UVB levels could have socio-economic implications for pet owners and those involved in animal husbandry, since they may be as or more significant than the direct effects of increased UVB exposure. In many cases the direct effects on animal health will be similar to those on the human population, especially with respect to those associated with the skin.

Effects are likely to vary between farmed animals, companion animals and wildlife. The more regimented the activity of the animal the less easily the animal can adapt to change. Effects are likely to vary amongst different sub-groups of terrestrial and aquatic animals. Within any of these different groupings the risk factors from increased UVB will vary according to the position of the animal within the food chain, the particular species, breed and its behaviour and ecology.

An estimate of the effects of increased UVB on animal health will require a knowledge of the epidemiology of any diseases likely to change under increased UVB, the dose-response relationship of any disease to UVB and the likely duration of exposure. Other than some work on laboratory animals, appropriate experimental studies on the possible effects of UVB on animals in natural environments have not been undertaken and few reports are available. At present, it is only possible to extrapolate on possible increases in the incidence of the direct effects of increased UVB on the basis of human and laboratory animal studies under analogous clinical conditions. Several of these problems may be less severe since many animals, especially farm animals and wildlife, have much shorter life-spans. The vulnerability of companion animals will be related to their owner's lifestyle.

7.2 Health

Skin

UVB induces short-term effects in animals such as irritation of the skin. This may be exacerbated by environmental chemicals obtained in food or by surface exposure to for example, alkaloids from pasture plants, or hygiene products used in farming such as disinfectants (Shearer *et al.*, 1992). Reports indicate that milk production is significantly affected in dairy cows with teats affected by sunburn. Outdoor pigs are also commonly affected by sunburn.

Longer term direct effects of higher UVB levels may include squamous cell carcinomas. These are common in cattle, sheep, horses, cats and dogs (Madewell & Theilen, 1987) and occur mostly on exposed skin unprotected by hair or pigment, including the eyes, nose, lips and ears. Certain breeds may be predisposed to such problems. Hereford cattle, (an important breed to the beef industry) lack scleral and eyelid pigment. Some 21% of Hereford cattle examined in the USA and Canada showed evidence of squamous cell carcinomas, mostly in older animals (Anderson & Badzioch, 1991). Recent examination of Hereford cattle in southern Chile suggests a similar rate of squamous cell carcinomas which does not appear to be related to recent changes in UV levels (Schein *et al.*, 1995). Squamous cell carcinomas are also more common in white and partly pigmented cats, especially those over 5 years old (Dorn *et al.*, 1971). White horses and sheep are also vulnerable to UVB. It is likely that the occurrence of squamous cell carcinomas will rise with increased exposure to UVB in those species and breeds already identified to be susceptible.

As with the human population, many of the potential effects may be minimised given suitable assessments of the risk and appropriate management action. Control of exposure to the increase in UVB projected by certain ozone depletion scenarios will be key. Cattle in areas with high solar intensity are provided with shade even when at pasture. Many cat owners already apply sun-blocks. The ability of animals to respond to increased UVB should however, not be underestimated, as thinly haired or hairless cats and dogs have been shown to adapt by tanning (Kimura & Doi, 1994). Management may be of limited value for animals with a genetic predisposition to UVB effects. Appropriate breeding programmes may contribute to medium term solutions for some species and breeds.

Eyes

Epidemiological studies have indicated that chronic progressive degeneration of the cornea, Überreiter's syndrome, appears in several breeds of dogs and is related to UVB exposure and more prevalent at higher altitude (Scatter *et al.*, 1977). UVB is also one of the factors in the development of a common disease of cattle, infectious keratoconjunctivitis caused by *Moraxella bovis*. The combination of exposure to UVB and infection by the bacterium significantly increases the severity of the lesions that develop, with cattle having little pigment near the eye being most susceptible. This condition is most common in young animals and so may have more of a general impact. It is likely that additional exposure to UVB may aggravate other chronic infections, such as latent viral infections.

Pathogens and vectors

UVB can have marked effects on microorganisms. High doses are used in sterilisation procedures, whilst low doses in natural radiation have little effect on bacteria. However, the survival of virus and fungi outside the host species may be decreased. A marginal increase in the rate of genetic mutations may be likely. However, it is unclear how bacterial growth and changes in virulence will be influenced without further study. Other pathogens are likely to be even less affected.

Many pathogens rely on mechanical and arthropod vectors e.g. insects and ticks for transmission between hosts. The consequences of changes to vector transmission for animals will be similar to those for human health but there are no adequate reports or estimates of effects. Any changes occurring in the UK will be likely to be minimal in response to increased UVB in comparison with those in tropical regions, simply from the differences in relative abundance of vectors. Other climatic changes, especially the effect of increased temperature and altered rainfall pattern, will probably overshadow the total effect of increased UVB on vectors.

Immune defences

The impact of increased UVB may lead to a variety of changes to the immune system, especially immunosuppression. This may result from direct effects on cellular immunity, for example, leucocytes respond to skin damage and a proliferative response will lead to immature and less capable cells in the circulatory system. Studies on laboratory rats (Goettsch *et al.*, 1994) indicate that even short-term exposure to UVB may seriously impede cellular defences to resist infection by bacteria, fungi or parasites, by making T-cells tolerant of the infectious agent. The consequences are likely to result in an increased risk of skin tumour development, failure to respond to infection and possibly a poor response towards any vaccine applied soon after exposure (de Gruijl & Forbes, 1995).

Other effects

Positive effects of increased UVB on animal health should not be overlooked. UVB irradiation has been shown to improve parts of the metabolism associated with bone turnover, perhaps related to increased vitamin D production. This leads to the suggestion that an increase in UVB could help prevent generalised bone disorders in high producing dairy cows (Philipov, 1992).

7.3 Husbandry and management

In managing farmed animals it is difficult to control a single environmental influence. Furthermore, it is difficult to elucidate the effect of increased UVB irradiation against other environmental variables, since high UVB conditions may also be linked to warmer temperatures, low rainfall and pasture availability. In this context, controlling exposure to UVB will be of minor importance and be related to the provision of shelter and foodstuff and any changes to these that relate to other aspects of climate change such as sun, rain and carbon dioxide. These may lead to changes in crops and thus supply of forage and other foodstuffs. The effect of nutritional changes may be predictable from farming systems in other countries although there are many unknowns. These include the effect of increased UVB on the quality of food for herbivores, which may be managed for farm animals but not for wildlife. Food may be of poorer quality, less palatable and of lower energy value if photosynthesis is impaired by UVB. In the short term, this may arise from an increase in the content of phenolic compounds in plant tissues, a plant defence response to UVB. In the longer term, there may be changes to the abundance and distribution of plant species. This may cause significant impacts to animal species which are dependent on specific plant species. In the short-term there may be effects on the nutritional and productive health of farmed animals and the well-being of wildlife. In the longer term production of food for farm animals may be managed but food sources for wildlife may be vulnerable to increased UVB.

Companion animals may respond to changes in their owners behaviour and hence reduce their exposure. They could also be expected to have some innate ability to minimise their exposure, even if the owners do not.

Overall, the consequences for wildlife can only be speculated upon. These include the possible rapid decline in amphibian populations in response to increased UVB (see Chapter 5). In general, wildlife will be more responsive to environmental changes since they are under greater selective pressure. The effect of increased UVB is likely to occur more slowly than the effect of changes in temperature or soil moisture. The indirect effects of elevated UV levels may be most significant and are likely to affect many species as a result of changes to nutrition, vectors of disease, responses to disease and the animal's ecosystem. Animals active in daylight and those in non-shaded environments will be most vulnerable to direct effects on health. Natural defence strategies, however, usually induce almost all species to use shade for protection against predators and hostile environments. Any general deleterious effects of UV may result from changes to the food chain which will affect all wild animals.

7.4 Timing of effect

The various responses of animals to increased UVB are likely to be observed over a number of timescales. In the short-term, the most obvious and more acute effects could occur to younger animals, providing the motivation to respond quickly to limit these impacts. Where effects are chronic and time-delayed, they may remain undetected and subsequent remedial action may be much less effective. Response to such effects will have to be planned on the basis of prediction and some degree of speculation. It is possible that the development of many chronic effects will be limited by the responses of individual animals or their populations; this is most likely in wildlife and where impacts may not even be recognisable. Planning to reduce the impact on farmed animals may be necessary, in these cases consideration should be given to selection of breeds and breeding policy.

The overall response must be to consider how to limit impacts by the selection of resistance to the effects of increased UVB and by developing management systems which will lower exposure and minimise adverse responses.

7.5 Conclusions

Determination of changes due to increased UVB and prediction of management responses necessary will require study. This could include experimental studies of effects on laboratory animals. Some are already being undertaken, however, many of these involve nocturnal animals and so do not provide good models for most animals at risk. Responses are already known to be species and breed variable. Long-term impacts can only be assessed if the present status with respect to vulnerability is known. It is becoming urgent that the sensitivity of various types of animals is assessed so that potential deleterious effects can be easily identified and appropriate actions taken to manage the risk.

8 Impacts on Materials

- **Materials made from natural or synthetic polymers are widely used in the construction, transport and agricultural industries. When exposed to sunlight (especially UVB) degradation can occur, this is exacerbated by moisture and high temperatures. At first, only the appearance may be affected but in the longer term loss of physical integrity and function can occur.**
- **It is now understood how most common polymers can be stabilised so as to provide acceptable durability to weathering. This technology can probably cope with currently anticipated UVB levels, at least in the relatively mild UK climate.**
- **The use of standardised natural and laboratory weathering tests helps to ensure acceptable performance in use.**

8.1 Introduction

All materials used out of doors are liable to be damaged by prolonged exposure to the elements; sunlight, rain, wind, temperature variations, frost and so on. However it is organic polymers, both of natural and synthetic origin, which are most at risk from the action of sunlight. It is the ultraviolet part of the solar spectrum which is the most damaging, especially the shortest wavelength, high energy UVB radiation. Should loss of ozone lead to significantly increased UVB radiation levels at the earth's surface, then the durability of organic polymer based materials will be adversely affected. The very wide variety of polymers in use, together with the many variations in formulation required to suit particular end uses make it difficult to generalise concerning the extent of problems which may arise. Even so, the nature of the problems and how to mitigate their effects are now well understood.

Organic polymers form the basis of many products used out of doors in the construction, transport, agricultural and leisure industries. They may take the form of plastics or rubber components, paints, textiles, gaskets or sealants. Examples of natural products include timber and thatch. Frequently, as with paints for example, such products are both protective and decorative in function. They are thus required to keep up good standards of appearance of the external surfaces. Colour fastness and retention of surface texture are then important in addition to maintenance of physical integrity.

8.2 Action of sunlight on organic polymers

Organic polymers are commonly formed of long chain structures built up of carbon, hydrogen and in many cases oxygen and nitrogen. Such materials are generally combustible and their degradation by sunlight bears some resemblance to combustion in slow motion. The processes involved in photo-oxidation are usually free radical chain reactions initiated by absorbed ultraviolet light and sustained by thermal reactions often involving oxygenated species such as peroxides. Eventually the original polymer network is reduced to small fragments. These can either evaporate, like carbon dioxide and water, or be washed away by rain.

The rate at which these reactions proceed, depends on the energy of the sunlight falling on the surface both in terms of the photo-active UV radiation and the overall energy in so far as this causes the temperature of the affected surface to rise. It is not easy to separate these two effects in practice. These weathering processes have been the subject of much research; a thorough overview is that by Davis and Sims (1983).

For photo-initiated reactions to occur it is necessary for light quanta in the sunlight both to be absorbed and to possess sufficient energy to cause specific chemical reactions as opposed to generalised heating. It is for this reason that the UVB content of sunlight is so significant despite the fact that it only comprises a small fraction of the total energy of sunlight. There is a higher probability that UVB radiation will be absorbed by the polymer itself and then initiate irreversible changes in chemical structure than for the more abundant, longer wavelength UVA. A review of the potential effects of enhanced UVB levels on polymeric materials has been published recently (UNEP, 1994).

A common scenario for photo-degradation is that UVB will contribute strongly to the initial stages in which chromophones absorbing longer wavelengths are produced, often leading to yellowing. These new chromophores in turn absorb and are bleached by longer wavelength, UVA, radiation. Eventually a steady state is set up as the exposed surface is more or less rapidly eroded.

For some polymers, oxidation processes can compete with degradation reactions which are specific to that polymer. An example is the very widely used polyvinylchloride (PVC). Experience has shown that PVC rooflights, if weathered in relatively cool, well ventilated situations tend to whiten. This indicates that photo-oxidation predominates. However installations such as barrel vault roofs, which if poorly ventilated, can become excessively hot in sunlight may cause PVC to undergo rapid degradation by loss of hydrogen chloride. This can cause blackening and, in the extreme, thermal distortion.

Exposure to prolonged outdoor weathering can have an adverse effect on the physical properties of polymers of all types. Most commonly this takes the form of increased brittleness reflected in a reduced ability to withstand impact damage. Materials such as rubbers, gaskets and sealants may lose resilience and hence the ability to withstand mechanical or thermal movement. Paints and surface coatings may become prone to cracking and flaking. Such behaviour is associated with irreversible changes in the polymer structure.

8.3 The stability to UV-radiation of commonly used polymers

Thermoplastic polymers

So far as stability to ultraviolet light is concerned, it is an advantage for a polymer if it does not absorb radiation present in sunlight. Examples of intrinsically stable polymers include polymethylmethacrylate (PMMA) and, especially, fluoropolymers such as polyvinylidene difluoride (PVDF). The latter is used to make exceptionally durable coatings for metals or fabrics. PMMA can benefit from the addition of UV-absorbing additives.

On the face of it, one might expect the bulk polymers of commerce, such as polyethylene, polypropylene and polyvinylchloride to show similar resistance to UV. It is true that these polymers have basic structures which do not absorb UV. However they also contain susceptible features, such as polymerisation initiation points or thermal and

Box 8

Polysulphone Film as a UVB monitor

It has been shown that polysulphone film is sensitive to radiation below 320 nm and this feature has been used to develop a simple means of monitoring UV (Davis, Dean and Diffey, 1976). When polysulphone film is exposed to radiation below about 320 nm, the initial broad absorption band extends towards the visible. This change is quantified by measuring the increase in absorbance of the film at 330 nm. This increase can be expressed in terms of equivalent dose of monochromatic radiation, 305 nm being used for calibration purposes.

Because of the similarity of the spectral response of polysulphone (Figure 8.1) to the erythema action spectrum of human skin, the film has been used as a personal UV dosimeter. Polysulphone has also been used to measure solar UV continuously in a number of locations and has been proved particularly useful to monitor the UVB output of artificial weathering light sources (Davis and Gardiner, 1987 and references therein). This method has been adopted as a British Standard (BS 2782 : Part 5 : Method 540C 1988). It is particularly useful in comparing different light sources, light filters and consistency of performance over time.

Figure 8.1

Activation spectrum of polysulphone film (compare with Fig. 6.1 the erythema action spectrum)

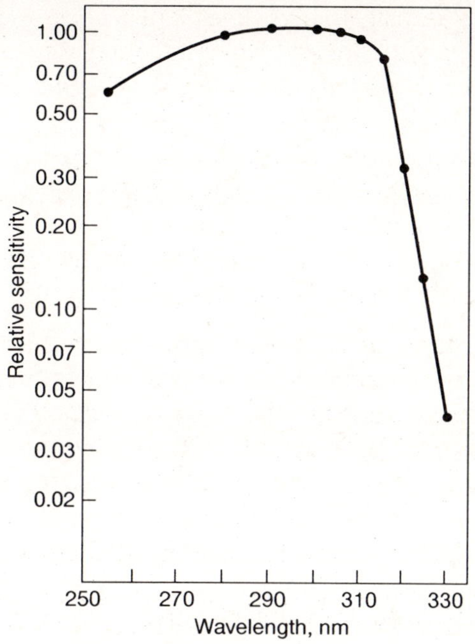

mechanical damage sustained during processing, from which photochemical reactions can develop. In addition, the polymers themselves are susceptible to attack by free radicals generated by photo-oxidation. Consequently, unstabilised polypropylene, for example, can be degraded very rapidly by sunlight. The role of stabilisers is crucial to the successful use of these polymers out of doors. Stabilisation of polymers is the subject of intensive study and development. Specific systems work best for each type of polymer. There is a very extensive literature; a useful review is available (Rabek, 1990).

Other polymer types, such as polycarbonates, have chemical structures which absorb UV and especially UVB, directly. Consequently, yellowing and surface erosion can develop in the medium to longer term. The performance of clear polycarbonate glazing can be significantly improved by incorporating a UV-absorbing surface layer at the time of manufacture.

Some UV-absorbing polymers have a very specific sensitivity to UVB at least in the early stages of exposure. An example is polysulphone which is used as a UVB monitor (Box 8). Some information is available concerning the spectral sensitivity of other types of polymer (UNEP, 1994). Substantial reductions in ozone levels could extend the spectrum of terrestrial UVB to lower wavelengths as well as increasing the overall intensity. This in turn might cause disproportionate damage to some materials, however there is a lack of information on the likely practical effects.

Thermosetting polymers

Thermosetting polymers are typically supplied as 2-part systems which react when combined to give a rigid, cross-linked structure. Once formed to shape they cannot be remelted in the way that thermoplastic materials can. Examples include unsaturated polyester resins, widely used as fillers or, reinforced with glass fibres, as GRP mouldings. Epoxy resins are widely used as adhesives, paints and fillers. Polyurethane resins commonly form the basis of flexible foams for furniture and rigid foams for thermal insulation. As commonly supplied such polymers are often UV-absorbing, aromatic chemicals and do not weather well when exposed to sunlight and UVB in particular. Therefore, 2-pack epoxy paints, for example, can be used as primer or undercoat in an aircraft paint system, providing high adhesion and chemical resistance. However it is necessary to use as weather resistant top coat paints such as 2-pack aliphatic polyurethanes, to give the required durability and standard of appearance in this demanding application.

Many commonly used polymers for application as paints and sealants for instance have quite complex structures and limited stability in sunlight. Alkyd resin based gloss paints, widely used for decorative purposes, are prone to lose gloss and "chalk" ie develop a powdery surface deposit after a few years in use. Such materials are liable to be degraded by longer wavelength UV in addition to UVB. However, their overall performance also depends on the quality of pigments, or other colouring agents used in their manufacture.

Biopolymers

Many biopolymer-based materials are used in situations where their performance may be adversely affected by UV degradation. For example, timber and wood-based products are widely employed as building components such as cladding, fencing and the like. Whilst such products can be protected from the worst effects of weathering by suitable treatments and surface coatings, this is at the expense of continued maintenance costs. Natural sources of textile fibres include cellulosics, such as cotton, and animal derived protein fibres, such as wool. These may be susceptible to UV degradation prior to harvesting, as well as in subsequent use. There seems to be a limited amount of information concerning the effects of increased UVB levels on these renewable materials (UNEP, 1994).

Light fastness of pigments and dyes

Most polymeric products are blended with pigments or dyes to provide the required colours; paints and fabrics are obvious examples. It is then a requirement that these colourants are adequately colourfast and do not promote degradation of the polymer matrix which contains them. Some pigments, such as carbon black commonly used with polyethylene, are stable and have useful protective properties. Others such as the widely used white pigment titanium dioxide (TiO_2) are strong UV absorbers but can cause polymer degradation at the polymer/pigment interface. The anatase form of TiO_2 is a strong UV pro-degradant. Specially coated versions of rutile TiO_2 are needed for durable white paints and plastics. Most other coloured pigments and dyes are strong absorbers of some parts of the UV and visible spectrum and so may not be particularly sensitive to UVB radiation compared to UVA and visible radiation.

The great variety of pigments and dyes makes it very difficult to generalise. Some very light fast pigments, notably strong yellows and reds derived from cadmium, are being withdrawn for environmental and safety reasons and may not easily be replaced.

Lightfast dyes are needed for fabrics, wall and floor coverings, etc which are exposed to sunlight through window glass. However, since window glass is an efficient absorber of UVB, this is not considered to be a problem in the present context. In special situations, such as museum displays, the protective action of the glass can be enhanced by application of thin polymer films transparent to light, but absorbing both UVA and UVB radiation.

Standardised test procedures for light exposure

The very wide range of polymeric materials in use, together with the added complexities of formulation variables and manufacturing procedures make it difficult to predict weathering performance from first principles. For this reason standardised weathering procedures have been developed for application to manufactured products. Internationally agreed test procedures are available for plastics, rubbers, paints and textiles involving exposure to natural daylight and daylight through window glass, together with a range of laboratory light sources.

Exposure to natural weathering of organic polymer-based materials is a very variable phenomenon depending on location, aspect, time of day, season of the year and so forth. In a temperate climate such as in the UK the effects of a long spell of hot sunny weather, as in the summer of 1995, can have a strong impact. Other climates may be more predictable. Exposure racks for natural weathering trials are commonly set up at 45° facing the equator. However for maximum exposure to UVB, which tends to be highly scattered, an angle of 10° to the horizontal facing the equator is more appropriate. Near horizontal surfaces illuminated by the whole sky will be most affected by increased UVB levels. This includes flat roofs, road surfaces, sports surfaces such as artificial turf and upper surfaces of vehicles and aircraft. As a consequence of this variability it would be quite difficult to detect with confidence the effects of modest increases in UVB radiation on performance in long term natural weathering trials. In addition, several year's exposure is usually needed to produce measurable effects.

Exposure to laboratory light sources offers an alternative approach which, at least in principle, is both quicker and more controlled than natural weathering for development work and performance based product standards. This also offers a means of assessing the sensitivity of materials to UVB radiation in the presence of longer wavelengths by appropriate choices of lamps or light filters (See BS 2782 : Method 540 D).

The use of xenon arcs as light sources (See BS 2782 : Method 540 E) comes nearest to matching the full solar spectrum, the UVB content being controlled by use of suitable filter combinations. Test pieces of various depths of colour will reach a similar range of surface temperatures to those induced by direct sunlight. In addition to temperature, factors such as water spray duration and frequency and relative humidity in dry periods can be kept reasonably constant. Light filters are generally designed to simulate either direct sunlight (Method A) or sunlight through window glass, (Method B) ie with most of the UVB removed. It is more difficult to fine tune control of the UVB intensity, however the levels achieved can be monitored using polysulphone films (See BS 2782 : Method 540 C).

An alternative approach is to use fluorescent lamps, most commonly available in the form of UV/ Condensation test cabinets (See BS 2782 : Method 540 F). UVA lamps with peak emission of 340 nm provide a close match to direct sunlight up to peak wavelength. UVA lamps with peak emission at 351 nm simulate sunlight through windowglass. Also available and widely used for highly accelerated tests are UVB lamps with peak emission at 313 nm.

These UVB lamps could be helpful in predicting the susceptibility of polymeric materials to shorter wavelength UVB which could reach the earth's surface should ozone levels be substantially reduced. However, all these lamp types are deficient in the longer wavelengths compared to sunlight. This can distort the balance of colour change in some cases. In addition this type of apparatus tends not to differentiate between surface temperatures reached by differently coloured test pieces.

The use of carbon arc lamps is long established but is now being phased out (See BS 2782 : Method 540 G). The light output is difficult to control and the spectral distributions differ appreciably from sunlight.

Photodegradable polymers

Finally it is worth noting that some polymer formulations are deliberately designed to degrade once used and left exposed to sunlight. Polypropylene is commonly used for this purpose and compounds which accelerate degradation can be added if required. Such products might benefit from higher UVB levels.

8.4 Indirect effects on polymer performance

Ground level ozone

If enhanced UVB radiation at ground level increases ozone formation (see Chapter 9) this could accelerate the degradation of natural rubber products and certain synthetic elastomers. The effects can be countered by the use of antiozonants and formulations can be strengthened if necessary.

Polymer foams and CFC's

Concern over the risk of enhanced UVB levels caused by ozone depletion has lead through the Montreal Protocol and subsequent agreements to a programme of phase out of CFC's and related compounds. CFC's were very widely used in the manufacture of polymer foams for thermal insulation and furnishings. The foam industry has had to extensively reformulate their products as a consequency, extruded polystyrene, polyurethane and phenolic foams all being affected. The long term performance in use of the new formulations has yet to be established.

Conclusions

The very wide variety of organic polymer based materials and their applications make it difficult to draw other than rather general conclusions regarding the effects of enhanced UVB radiation in sunlight. Taking the most extreme case of a polymer exclusively sensitive to UVB radiation, it is expected that the rate of degradation would be increased in proportion to the increase in UVB. Other systems would be proportionately less affected.

The technology to improve the stability of most polymers to UV degradation already exists so that the durability of susceptible materials can be improved if necessary, albeit at some cost.

Failing this, some increase in maintenance and replacement costs is to be expected. Because of the great variety of materials and end-uses involved it would be a complex task to try to estimate this increase in costs. A rough estimate of the worst case could be taken as being in proportion to the increase in UVB levels. However other factors, such as temperature, especially maximum air temperatures in periods of prolonged sunshine are also significant.

British Standards

BS 903:- Physical testing of rubber:-

Part A53 : 1989 Methods for exposure to natural weathering

Part A54 : 1989 Methods for exposure to artificial light

BS 1006:- Textiles - Tests for colour fastness:-

(= ISO 105)

Part B03 : 1994 Colour fastness to weathering: outdoor exposure

Part B04 : 1994 Colour fastness to artificial weathering

-Xenon arc fading lamp test

BS 2782:- Methods of testing plastics

Method 540C : 1988 Determination of ultraviolet radiation intensity using polysulphone film

Method 540D : 1995 Methods of exposure to laboratory light sources; General guidance (= ISO 4892-1 : 1994)

Method 540E : 1995 Xenon arc sources (= ISO 4892-2 : 1994)

Method 540F : 1995 Fluorescent lamps (= ISO 4892-3 : 1994)

Method 540G : 1995 Open-flame carbon-arc lamps (= ISO 4892-4 : 1994)

Method 550A : 1981 Methods of exposure to natural weathering (to be replaced by ISO 877 : 1994)

BS 3900:- Methods of test for paints

Part F5 : 1972 Notes for guidance on the conduct of natural weathering tests

Part F14 : 1995 Determination of resistance to artificial weathering by exposure to filtered xenon arc radiation. (= ISO 11341 : 1994)

9. Air Quality

- **Increases in UV could lead to increases in ozone production over the cities and some downwind areas. Increased exceedance of ozone air quality standards in urban areas is anticipated. Rural air quality is likely to be largely unaffected by UV increases.**
- **The increase in UV resulting from a 5% column ozone reduction increases hydroxyl (OH) radical concentrations by about 4%. In the free troposphere, well above the cities, ozone is depleted, resulting in a 1 to 2% reduction in column integrated tropospheric ozone. Changes are also expected in the tropospheric distribution of HO_2 and hydrogen peroxide.**
- **Methane (CH_4) is an important greenhouse gas which has been increasing in the atmosphere due to human activities. The expected increase in OH due to the observed global ozone reduction is expected to have reduced CH_4 concentrations by about 4%. However, taking into account the continued release of CH_4, the observed ozone reduction cannot account for all of the observed decrease in the CH_4 growth rate.**
- **UV at the ground is reduced by the presence of particulate matter and aerosols in the lower atmosphere. Strategies to control particulate pollution will result in increased UV levels.**

9.1. Introduction

The term air quality covers a wide range of phenomena occurring over a wide range of spatial scales. It includes the formation of local, urban scale pollution events which are largely driven by the competition between vertical dispersion and the emissions from human activities. It also includes the oxidation and removal of a wide range of trace gases, which occur throughout the troposphere from the global scale atmospheric circulation in both hemispheres. Both these processes share the same feature, that the chemical climates which drive them, are controlled in part by the intensity and wavelength distribution of solar ultraviolet radiation. The photochemical destruction of photochemically reactive molecules, such as ozone and formaldehyde (HCHO), is a major stimulus to tropospheric chemistry and to the deterioration of tropospheric air quality.

Tropospheric chemistry, like many features of tropospheric behaviour, is driven by processes occurring at its boundaries, i.e. stratospheric processes (see Chapter 2) and processes occurring at the ground. Throughout the troposphere, the intensity and wavelength distribution of ultraviolet radiation is highly variable and is influenced by the presence of clouds, aerosol particles and the nature of the underlying surface and its reflecting properties. All things being equal, the less ozone there is in the stratosphere, the higher the intensity of the solar ultraviolet radiation in a particular wavelength range in the troposphere and the more this wavelength distribution extends to low wavelengths.

The suggestion has been made that the increase in tropospheric ozone concentrations could counteract the effect of stratospheric ozone depletion on the amount of ultraviolet radiation reaching the Earth's surface (Bruhl and Crutzen, 1989). Bruhl and Crutzen (1989) illustrate their suggestions using data from the Hohenpeissenberg observatory in Bavaria for which continuous records of tropospheric ozone and column ozone exist since 1968.

A detailed review of ozone air quality in the United Kingdom is given in the report of the Photochemical Oxidants Review Group (PORG, 1993). In this section attention is directed solely at the effect of changes in ultraviolet. The first comprehensive assessment of the impact of UV changes on tropospheric air quality was published as part of the UNEP assessment of the environmental impacts of stratospheric ozone depletion (Gery, 1989). This review addressed:

- the mechanism of the chemical response to increased UV,
- the effects on remote and rural air quality,
- the effects of increased UV or urban air quality,
- combined affects with changing global climate.

The issues and conclusions stand intact despite the period of intervening research covering the last six years or so. In the following sections, two of the issues are followed up in some further detail and recent research results are reviewed. Reference is also made to the second comprehensive review of the impact of UV changes on troposheric air quality published by UNEP as part of their 1994 assessment report (Tang and Madronich, 1994).

9.2. Ultraviolet Changes and tropospheric chemistry

Current understanding of tropospheric photochemistry focuses attention on the two free radical species, hydroxyl (OH) and hydroperoxy (HO_2), and their interconversion reactions. Daylight concentrations of these free radicals are controlled by the trace gas concentrations, their chemical reactivities and the intensity of the solar ultraviolet radiation, see Box 9. The concentration of hydroxyl radicals, so formed, reaches a peak concentration of only a fraction of a ppt (1 ppt = 1 part in 10^{12} air molecules). This is nevertheless enough to remove 400 megatonnes of methane (CH_4) annually, 2 gigatonnes of carbon monoxide (CO), another gigatonne of hydrocarbons from both natural and man-made sources. OH also removes in part smaller quantities of almost any trace gas with sufficient reactivity, including the CFC replacements hydrochlorofluorocarbons (HCFCs) but not the CFCs themselves, which are totally inert in the troposphere and can be carried upwards ultimately to destroy stratospheric ozone (Chapter 2).

Fuglestvedt *et al.* (1994) examined the impact of the reduction in ozone column that has been reported and modelled from 1970 to 1980 and from 1970 through to 1990. Relative to 1970, they calculate a 4.5% reduction in the global annually averaged ozone column due to human activities. Table 9.1 presents the percentage changes in the peak and globally averaged photolysis rate coefficients (see Box 9) for some important photochemical reactions in the troposphere. The percentage increases in the photolysis rate for the most important tropospheric photolytic process:

$$O_3 + hv \dashrightarrow O(1D) + O_2,$$

are substantially greater than the corresponding reductions in ozone column. Apart from this process and the photolysis of nitric acid, the percentage increases in photolysis rate coefficients are smaller than the decreases in the ozone column (Fuglestvedt *et al.*, 1994). Hence the main impact of UV changes will be on O(1D) (and hence OH) and nitrogen oxide concentrations.

Table 9.2 shows the resulting changes in the mean tropospheric concentrations of hydroxyl radicals, methane and ozone tropospheric abundances. In response to the ozone column depletion, UV intensities increase, photolysis rates increase, mean tropospheric hydroxyl radical concentrations increase and methane and tropospheric ozone abundances decrease. The changes in tropospheric ozone are the opposite sign and somewhat smaller than those in the mean tropospheric hydroxyl radical concentration, which is in turn smaller than that in the photolysis rate coefficients (Fuglestvedt *et al.*, 1994).

It is concluded from current models that enhanced UV will increase the OH, HO_2 and H_2O_2 concentrations in the troposphere. The extent of any increase remains uncertain with model results being sensitive to model formulation. The results quoted above are meant to be illustrative and are from a state-of-the-art global three-dimensional chemistry-transport model (Fuglestvedt *et al.*, 1994). Results from a simple zero-dimensional chemistry model (Liu and Trainer, 1988) and a one-dimensional model for different chemical regimes (Thompson *et al.*, 1991) are quantitatively different (Tang and Madronich, 1994) but are not discussed here because of the inherent weaknesses of limited dimension models.

Box 9

Photochemistry basics

For a reaction such as **$OH + CH_4$ --> $H_2O + CH_3$**
The rate of change of CH_4 is given by **$d[CH_4]/dt = -k[OH][CH_4]$**
where square brackets denote concentrations and k is the reaction rate which depends on the temperature and for some reactions pressure as well.

For reactions involving sunshine (photolytic reactions) such as **O_3 + hv --> O(1D)** + O_2
hv denotes a photon and the rate of change of O_3 is given by **$d[O_3]/dt = -J[O_3]$**
where J is the photolysis rate, which depends on the amount of UV reaching the troposphere and hence on the stratospheric ozone column.

The Fast Photochemical Balance of the sunlit troposphere.

Free Radical Production Reactions:

O_3 + hv --> O(1D) + O_2
O(1D) + H_2O --> OH + OH
HCHO + O_2 + hv --> HO_2 + HCO

Free Radical Loss Processes:

$HO_2 + HO_2$ --> $H_2O_2 + O_2$
$OH + HO_2$ --> $H_2O + O_2$
$OH + NO_2$ --> HNO_3
OH to HO_2 Interconversion Processes:
$OH + CO + O_2$ --> $HO_2 + CO_2$
$OH + CH_4 + O_2$ --> $H_2O + CH_3O_2$
$OH + O_3$ --> $HO_2 + O_2$
$OH + H_2 + O_2$ --> $HO_2 + H_2O$
OH + isoprene --> HOC_5H8O_2
HO_2 to OH Interconversion Processes:
HO_2 + NO --> OH + NO_2
$HO_2 + O_3$ --> $OH + O_2 + O_2$

Ozone Production and Loss Processes:

HO_2 + NO --> OH + NO_2
CH_3O_2 + NO +O_2 --> $NO_2 + HO_2$ + HCHO

followed by:

NO_2 + hv --> NO + O
O + O_2 --> O_3
O_3 + hv --> O(1D) + O_2
$O_3 + HO_2$ --> $OH + O_2 + O_2$
O_3 + OH --> $HO_2 + O_2$
O_3 --> dry deposition
O_3 + NO --> $NO_2 + O_2$
$O_3 + NO_2$ --> $NO_3 + O_2$
$NO_2 + NO_3$ --> N_2O_5
$N_2O_5 + H_2O$ --> ${}_2HNO_3$

Table 9.1.

Percentage changes in the photolysis rates of a number of important photochemical processes in the troposphere due to a 4.5% decrease in ozone column, after Fuglestvedt *et al.* (1994).

Process	Percentage change in photolysis rate for June 50°N, noon	Percentage change in global annually averaged photolysis rate
O_3 + hv --> O(1D) + O_2	12.7	6.3
NO_2 + hv --> NO + O	0.3	0.2
H_2O_2 + hv --> OH + OH	2.6	1.6
HNO_3 + hv --> OH + NO	2 7.3	4.0
HCHO + hv --> H + HCO	3.1	2.1
HCHO + hv --> H_2 + CO	1.0	0.7

9.3. Tropospheric chemistry and the greenhouse effect

Methane concentrations exert a greenhouse effect in an analogous manner to carbon dioxide, but with a fraction of the intensity. Over the last few decades, global surface methane concentrations have been rising steadily by up to 10-20 ppb/year (Khalil and Rasmussen, 1994) due to man-made sources. This annual rate of increase has slowed measurably during the 1980s, and during the early 1990s it almost stopped altogether (Dlugokencky *et al.*, 1994), thereby reducing the relative contribution played by methane to global warming.

A possible explanation for the slowing down in the annual rate of increase in global methane concentrations is that stratospheric ozone depletion has increased tropospheric solar ultraviolet radiation with a corresponding increase in tropospheric OH (Bekki *et al.*, 1994). Modelling studies have shown that stratospheric ozone depletion may have accounted for the slow decline in the global methane growth rate before 1985. In the late 1980s, the slight recovery in column ozone may have prevented an even more pronounced decrease in the methane growth rate. Furthermore, the record low ozone column values in 1992 may have accounted for almost one half of the marked decrease in the methane growth rate also observed in that year.

The study of Bekki *et al.* (1994) shows that changes in the carbon monoxide abundances would also result from the impact of stratospheric ozone depletion, following the methane changes, step by step. The atmospheric abundances of carbon monoxide have certainly increased with the methane during the 1970s and early 1980s. However, the growth rate in the global mean carbon monoxide concentrations started to decline earlier than found with methane and changed sign between 1987 and 1990 (Novelli *et al.*, 1994). This is long before the period of peak ozone depletion and suggests that other factors besides stratospheric ozone depletion may be involved with both carbon monoxide and methane (Bekki *et al.*, 1994).

9.4. Ultraviolet Changes and Urban Air Quality

Photochemical smog was first detected in the Los Angeles basin in the 1950s (Haagen-Smit *et al.*, 1953). Subsequently as air quality monitoring networks have expanded, elevated ozone concentrations have been observed downwind of almost all of the major urban and industrial centres during summertime. Europe is no exception (Nordic Council of Ministers, 1991) and during most summers, elevated ozone concentrations are found, exceeding air quality standards set to protect human health and sensitive crops and plants in natural ecosytems.

The mechanism of photochemical ozone formation has been established through a combination of laboratory investigation, field investigation, air quality monitoring and computer modelling (Leighton, 1961; Demerjian *et al.*, 1974; Atkinson, 1990). Two regimes have been identified in which ozone formation is hydrocarbons or nitrogen oxides (NO_x) limited. In the hydrocarbon limited regime in the hydrocarbon and NO_x source areas over the cities, there is adequate NO_x to react almost

Table 9.2

Percentage changes in the global average values of ozone column, photolysis rates, tropospheric mean hydroxyl concentrations, for a 4.5% decrease in ozone column, after Fuglestvedt *et al.* (1994).

Parameter	Percentage change
ozone column	-4.5
J: O_3 + hv --> O(1D) + O_2	+6.3
tropospheric mean OH	+4.0
global methane abundance	-3.5
tropospheric ozone column	-1.4

quantitatively with the hydrocarbon products to produce ozone. In the NO_x-limited regime in the countryside, there is an inadequate supply of NO_x and this limits ozone production. NO_x has a short atmospheric lifetime because it is rapidly converted to nitric acid which takes no further part in the photochemistry. Once the NO_x becomes depleted, all ozone production ceases because the hydrocarbon products react with other peroxy radicals, short-circuiting the ozone production reactions.

All other things being equal, increases in solar ultraviolet radiation will increase the supply of hydroxyl radicals and will stimulate ozone production in those atmospheric regions which are in the hydrocarbons limited regime. In the NO_x limited regime, near the surface there should be little influence of solar ultraviolet changes on ozone production.

In looking at ozone control strategies for Europe, it is possible to divide Europe into those regions which are likely to be in the hydrocarbons and NO_x limited regimes. The high emissions areas of north west Europe in the Netherlands, Belgium, Germany and the United Kingdom are likely to be hydrocarbons limited and show an increased ozone production with increasing solar ultraviolet radiation. The remote fringes of Europe covering France, Ireland, Scotland and Scandinavia are likely to be NO_x-limited and local production in these countries is likely to be unaffected by UV changes. However, long-range movement of air across national boundaries will give rise to local ozone amounts which reflect upwind conditions which may be hydrocarbons limited.

In countries throughout the world, emission control strategies are being planned and implemented to limit the production of urban ozone. These strategies are based on the control of volatile organic compounds (VOCs) and NO_x emissions. Until recently, no change in stratospheric ozone had been expected and therefore pollution control strategies have not addressed futures with increased UV. For the cities in the USA, additional unanticipated VOC emission controls of up to 13% or so would be needed to maintain present ozone air quality with a 20% reduction in column ozone (Gery, 1989). De Leeuw and Van Rheineck Leysius (1991) have similarly estimated that given constant European NO_x and VOC emissions, urban air quality exceedances for ozone will be much more frequent in European urban areas if UV levels increase.

In summary, it is likely that increased solar ultraviolet radiation will lead to a deterioration in air quality with respect to ozone in cities and in their downwind areas. Nevertheless, this deterioration should be seen against the background of the considerable year-to-year variations in the intensity and frequency of elevated ozone concentrations, dictated by variations in meteorological conditions.

9.5. Particulate Matter, Aerosols and Ultraviolet changes.

Particulate matter and aerosols are microscopic particles and droplets which, because of their small size, are suspended in the atmosphere by vertical wind motions and carried from place to place by horizontal winds. After solar elevation, cloud and column ozone, particulate matter is the major factor in determining surface UV (see Chapter 2). Tropospheric pollution on the local scale can decrease UV, particularly during hazy, sunny weather. Particulate matter and aerosols decrease the direct solar beam by scattering UV radiation, thus changing the ratio between the direct and scattered beam. For an overhead sun with pollution in the lowest levels of the atmosphere the total UV incident on a horizontal surface is only expected to be slightly affected because the reduction in the direct beam is slightly compensated by an increase in the scattered beam. On the other hand for the solar elevations which occur for much of the time in the UK and northern Europe, the direct beam may be considerably reduced resulting in a disproportionate decrease in UV. It follows that strategies to control particulate matter and aerosol pollution could have a bigger effect on UV levels than stratospheric ozone depletion.

10 Socio-economic factors and impacts

- **Increases in UV-related diseases suggest that population exposure to solar ultraviolet radiation has increased. The dominant influence has probably been changes in behaviour. Changes in levels of ultraviolet radiation in the environment have so far probably played little part in the increase.**
- **Factors leading to an increase in exposure of the population to solar UV radiation include the growth in foreign holidays, increased participation in outdoor leisure activities, changes in clothing, the continuing popularity of sunbathing and a sun-tanned appearance.**
- **There is potential for any human health effects of increases in levels of UV radiation in the environment to be mitigated by changes towards more cautious behaviour. However, there are some important obstacles to such changes.**
- **Socio-economic implications of possible health impacts include costs of treatment, costs of programmes of prevention, the value of lives that are lost and the effects on the tourism and leisure industry of any change in behaviour resulting from perceptions of increased risks.**
- **There might also be economic implications arising from the effects of enhanced UV on agriculture, forests, fishing and on materials.**

10.1 Introduction

Previous chapters have shown that increases in the amount of solar UV radiation in the environment could have impacts in areas such as human health, agriculture and natural marine and terrestrial ecosystems. In turn, these UV induced changes might have a number of important socio-economic impacts and it is one of the tasks of this chapter to outline what these might be. Since these socio-economic impacts lie at the end of a long chain of causality it is inevitable that this involves many uncertainties. It has already been stressed that it is uncertain what may be the extent and time-course of future levels of UV radiation. The biological, physical and chemical impacts this might have are also in doubt. Further complexity is introduced by the fact that the consequences of such impacts for human welfare are likely to be vary in different social and economic settings.

In some areas the interplay between environmental changes and socio-economic factors can be extremely complex. A good example of this is the possible effects of increased levels of UV on human health. Any such impacts would be the result of increases in the dose of UV radiation received by the population. However, such doses depend not only on the amount of UV radiation in the environment but also on the patterns of human behaviour that influence exposure. Therefore, social and economic factors are likely to have an important bearing on the scale of the possible health impacts from an increase in UV in the environment. This chapter begins by considering the socio-economic factors that might influence the impacts of UV changes in other areas, particularly human health. It then goes on to consider the socio-economic impacts that might arise from changes in other systems.

10.2 Social and economic factors affecting human UV doses

Developed countries with fair-skinned populations have seen a large and sustained increase in skin cancer in recent decades. Figure 10.1 shows that England and Wales has followed this general trend with a substantial increase in melanoma deaths during the past four decades (although changes in

Figure 10.1

Melanoma deaths in England and Wales, 1949 to 1993

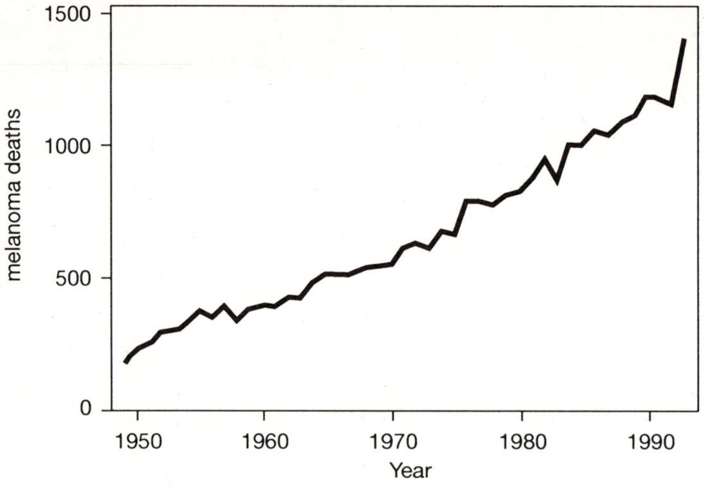

Source: OPCS Mortality Statistics : Cause

population size, age structure and case fatality rates may have also influenced trends). In the light of evidence linking ultraviolet radiation and skin cancer this has led to the conclusion that such increases have probably been the result of greater population exposure to the sun (Glass and Hoover, 1989). This has not been the result of changes in solar ultraviolet radiation flux which have changed little in such countries over this time scale. Instead, it is more likely that the increase in skin cancers has been the result of socio-economic factors involving changes in lifestyle and behaviour. Several factors may have been important including changes in occupational structure, the growing popularity of holidays abroad, increased participation in outdoor leisure activities, changes in clothing fashions, the popularity of sunbathing and a suntanned appearance. A factor underlying several of these trends has been the growing affluence of the population of countries such as the United Kingdom which may have led to more intermittent intense exposure to UV. The link with affluence can be seen in Table 10.1 which shows that proportional registration ratios for melanoma in England and Wales rise strongly with social class and similar associations have been found elsewhere. More recently, there are indications in the USA of a downward trend in melanoma mortality rates in younger people which may reflect changes in behaviour towards reduced UV exposure, especially during childhood (Scotto *et al.*, 1991)

Occupation

Outdoor occupations such as farming, forestry, fishing and some construction work can involve much greater exposure to sunlight than is the case for the rest of the population (Diffey, 1987). There is evidence that this is associated with a higher risk of non-melanoma skin cancers, particularly squamous cell carcinomas and more equivocally for basal cell carcinomas (World Health Organization, 1994). However, the number of people working in such occupations has decreased and therefore occupational risks cannot account for the rise in the incidence of skin cancers that has been observed. Nor does it appear that outdoor occupations are associated with a higher risk of malignant melanoma, with some studies actually showing a significantly lower risk for outdoor workers (Elwood, 1992). For malignant melanoma it has been suggested that the main risk is from intermittent, intense expose to ultraviolet radiation, not from total cumulative dose and, in this case, recreational habits seem to be particularly important (Elwood,1992).

Holidays abroad

It has been calculated that for a representative north European indoor worker a two-week summer sunbathing holiday in the Mediterranean area might be associated with more than a doubling of erythemally-weighted solar UV radiation doses (Diffey, 1987). Changes in the numbers of people taking part in such holidays could therefore be an important factor leading to an increase in solar UV

Table 10.1

Standardised proportional registration ratios for malignant melanoma of the skin for males aged 15-74 in England and Wales, 1974-84

Social Class		Proportional registration ratio
I	Professional	171
II	Intermediate	133
IIIn	Skilled non-manual	115
IIIm	Skilled manual	69
IV	Partly skilled	67
V	Unskilled	47

Source: OPCS Cancer Statistics

doses and any associated risks (Acheson, 1986). Furthermore, such exposures exemplify the characteristics of intermittency and intensity that appear to carry particular risk. Figure 10.2 shows that since the early 1970s there has been a very large increase in the number of holidays taken abroad by the British population with package tours to destinations such as the Mediterranean countries being a particularly important component of this trend. In recent years there are indications of a levelling off in the increase but the growing popularity of holidays to long-haul destinations at low latitudes means that UV exposures associated with holidays abroad may have continued to increase. Table 10.2 shows that there are marked social class differences with professional/ managerial groups showing the highest rates of participation in holidays abroad, although there is evidence that these class differences have lessened over time. If such activities do increase the risk of malignant melanoma their earlier and greater involvement in foreign holidays may account for some of the social class gradient in melanoma shown in Table 10.1.

Outdoor recreation

Although holidays abroad may involve substantial increases in solar UV doses there is evidence that exposures in other settings are also important. In particular, a number of case-control studies have shown that other outdoor recreational activities are associated with an elevated risk of malignant melanoma (Elwood, 1992). Part of the increase in skin cancers may therefore reflect increasing rates of participation in outdoor recreational activities such as boating, golf, etc. There are also some striking social class differences (see Table 10.3) with higher rates of participation by those in the professional and managerial groups paralleling the social class gradient in incidence of malignant melanoma.

Clothing and fashion

Changes in clothing fashions have the potential to change the amount of the body that is exposed to potentially damaging ultraviolet radiation. Magnus (1991) has shown that the fashion for topless sunbathing was followed by an increase in malignant melanoma on the female breast in Norway. Gender differences in customary clothing may be a factor in the higher incidence of malignant melanoma for females than for males in England and Wales. In particular, melanomas of the lower limbs are substantially more common in females than in males, perhaps reflecting gender differences in the likelihood of the legs being exposed by the usual style of clothing.

In summary, there is considerable evidence that socio-economic factors exert a major influence on the population's exposure to solar ultraviolet radiation and the risks associated with it. It seems likely that trends such as the growing popularity of holidays abroad and greater participation in outdoor leisure activities have had an important influence on UV doses and the health risks associated with them. The optimistic corollary of this conclusion is that changes in behaviour could offer an effective means of reducing some of the health risks associated with current levels of UV in the environment as well as from any enhanced levels in the future. This raises the question of socio-economic factors that might influence the potential for prevention of UV health impacts.

Figure 10.2

Number of holidays (millions) of 4+ nights taken by the British population, 1970 to 1990

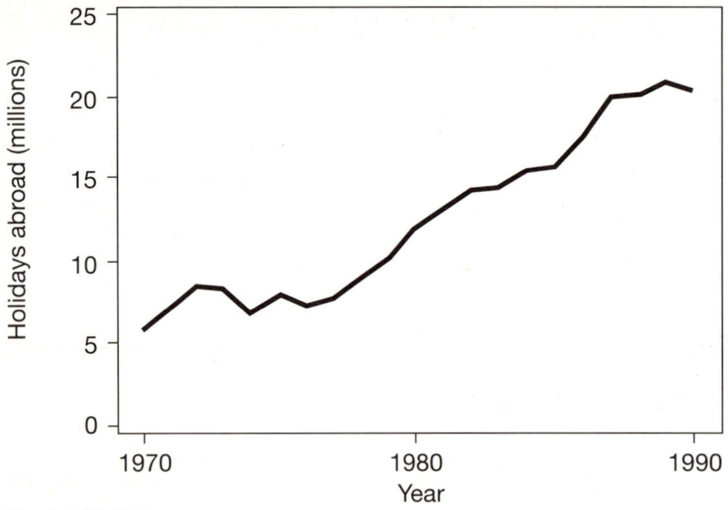

Source: Digest of Tourist Statistics

10.3 Preventing health risks by changing behaviour

If patterns of behaviour were to follow their current trends, higher levels of solar ultraviolet radiation in the environment could lead to an increase in the incidence of diseases such as skin cancer. However, this is not inevitable. Even if UV levels rise, changes in behaviour offer the potential for reducing risks. The health impacts of enhanced levels of UV in the environment could be mitigated either by encouraging changes in behaviour aimed at reducing exposure to harmful UV radiation (primary prevention) or by minimising the harmful effects of UV doses by early detection and effective treatment of diseases resulting from such exposure (secondary prevention).

This task is unlikely to be easy. Behaviour in respect of exposure to the sun is influenced by positive associations between a sun-tanned appearance and images of beauty, sexual attractiveness, wealth and health. Such attitudes are undoubtedly supported by the imagery of advertising by the tourism and leisure industry. However, they have not always been the same and in the Eighteenth and Nineteenth Century a pale complexion carried equally positive overtones. This points to the conclusion that the present attitude is not inevitable but is socially constructed and could change. A barrier to change is that there is a tendency for people to under-estimate risks (such as exposure to the sun) that are familiar and under personal control, particularly when they are associated with activities that are pleasurable (British Medical

Table 10.2

Index of participation in holidays abroad by the British population by social group (all social groups=100), 1970 and 1990.

	Holiday abroad index (all classes=100)	
Social Group	**1970**	**1990**
AB Professional/managerial	221	176
C1 Clerical/supervisory	138	123
C2 Skilled manual	78	89
DE Unskilled/pensioners etc.	43	53

Source: Digest of Tourist Statistics

Table 10.3

Participation in outdoor leisure activities by socio-economic group, Great Britain 1990

	% participating in previous 12 months					
	Walking	Outdoor	Golf	Tennis	Cricket	Sailing
Professional	81	39	24	15	10	9
Employers & managers	72	31	22	10	5	5
Other non-manual	69	24	11	8	3	3
Skilled manual	61	19	12	4	4	2
Semi-skilled	55	13	6	4	2	1
Unskilled	49	8	4	2	1	0

Source: General Household Survey

Association, 1987). On the other hand, unfamiliar risks that are imposed on people tend to be over-estimated. If the health risks of UV exposure were increasingly seen to be associated with a pollution threat to the global environment this might encourage a change to more cautious behaviour.

10.4 Socio-economic impacts of enhanced UV levels

Socio-economic implications of health impacts

It has been concluded in Chapter 6 that increased levels of solar ultraviolet radiation might lead to increases in skin cancers, eye cataracts and changes in the immune system that could alter the course of cancer or infectious disease. Such increases would be likely, in their turn, to have a number of socio-economic implications.

Costs of care

One of the most obvious of these is the cost of treatment of such UV related diseases. For conditions such as non-melanoma skin cancer and cataracts treatment is usually relatively simple and inexpensive. However, the costs of treating such conditions for the population as a whole might be substantial because of the large number of cases. Melanomas are much less common but the costs of treatment per case are likely to be higher, particularly where patients are diagnosed when the disease is at a more advanced stage. There may also be significant costs of social care and support in situations such as untreated cases of cataracts which may involve long term disability.

Value of life

In England and Wales in 1993 there were 1397 deaths from cutaneous malignant melanoma and 362 deaths from non-melanoma skin cancers. Although the primary impact here is clearly one of human tragedy there have also been attempts to consider such effects in economic terms (Pearce *et al.*, 1992). Deaths may involve a loss of future productive output which represents an economic loss. Over and above this it can be argued that society is willing to pay to avoid deaths. The approach of placing a value on "statistical lives" is routinely applied by the Department of Transport (DOT) when evaluating transport investments which might alter accident risks. A recent review (Pearce *et al.*, 1992) has proposed that a value of about £2 million would be appropriate for such assessments. This figure draws on many reported studies, many of which are US based. The DOT's value of statistical life is given as £826,600 (1996, value). However, in particular cases values might be affected by factors such as the age at which death occurs, with deaths at younger ages leading to greater losses of expected years of life and of future productive potential. In the case of UV-related disease it is relevant that, whereas most non-melanoma skin cancer deaths are amongst the elderly, significant numbers of melanoma deaths occur in young and middle aged adults. In England and Wales in 1993 there were 207 melanoma deaths in individuals aged less than 45 years and a further 468 in the 45 to 74 age group.

Cost of preventative measures

It has been stressed above that much of any adverse impact of enhanced UV could be avoided. But this might involve significant costs. There is a large (about £100 million per year in the UK) and growing market for sunscreens, but these are expensive, particularly for the formulations providing higher sun protection factors. In addition to their cost there is concern that the use of sunscreens might encourage users to spend longer in the sun, which could account for the elevated risk of melanoma found in users of sunscreens in two recent studies (Westerdahl *et al.*, 1995; Autier *et al.*, 1995). Therefore, while they may protect against erythema it is less certain that sunscreens provide the same level of protection against other effects such as the development of melanomas or basal cell carcinomas (World Health Organization, 1994). Campaigns aimed at changing behaviour also involve economic costs as would any policies aimed at secondary prevention, for example by encouraging earlier diagnosis of melanomas by screening. In industries such as agriculture involving outdoor work more attention may need to be paid to the risks posed by UV exposure. However, the effectiveness of simple measures such as changes in clothing and headwear suggest that major changes in work practices are unlikely to be necessary.

Effects on the market for tourism and leisure

As well as such direct costs any widespread changes in behaviour in respect of exposure to the sun might have substantial indirect economic effects. In particular, any increase in public concern about the dangers of sun exposure could affect the market for tourism and leisure activities. There might be a reduction in demand for beach based holidays and activities, especially those in destinations such as the Mediterranean countries and other low-latitude locations. Conversely, the market position for other types of tourist and leisure activities which are less based on the attractions of exposure to the sun could improve. This might involve an improvement in the competitive position of the tourist industry in Britain and other northern European countries. Such changes could have implications not only for the profitability of the companies concerned but also for employment and for the balance of payments

Agriculture and forestry

There is evidence that increased exposure to solar ultraviolet radiation can affect some agricultural crops and animals. Since profitability depends critically on yields there is an obvious potential for enhanced UV to have an adverse economic impact on agriculture. However, evidence is sparse on the UV sensitivity of arable crops that are commercially important in the United Kingdom and there is a shortage of data on UV impacts under local conditions. The present data are therefore inadequate to draw any firm conclusions on any impact on agricultural yields and on the economics of farming. Even if further research were to show that currently important crops were vulnerable this would not necessarily imply a major adverse economic impact. It is known that cultivars of the same species can exhibit large variations in susceptibility to UV damage. There is therefore a potential for any UV impacts to be mitigated by the substitution of vulnerable cultivars by those that are less sensitive to UV. If this were to happen, the economic impact would be confined to any remaining differences in yields and in costs of production between current crops and their less UV sensitive alternatives. Similarly, while enhanced UV might have short-term effects on the nutritional and productive health of farmed animals, management changes in the longer term are likely to reduce adverse impacts.

Evidence on which to base an assessment of UV impacts on forestry is even more sparse than it is for agriculture. However, there is evidence that the biomass of commercially important temperate-latitude tree species can be reduced by enhanced exposure to UV radiation. An adverse impact on yields and on the economics of forestry is therefore a possibility. The long production cycle in forestry also means that there is less potential to avoid adverse economic impacts by switching to more UV tolerant tree types.

Fisheries

At a global scale about a third of human consumption of animal protein comes from the sea, with a higher dependence in many developing countries. Any reduction in marine productivity as a result of enhanced exposure to UV radiation could have serious effects on human diets, particularly for vulnerable populations in the third world. The greatest concern is that increased exposure of phytoplankton to UV radiation could lead to a loss of primary biomass productivity which might result in lower fish catches. Within the United Kingdom this factor might add to several others that are reducing catches and increasing the economic difficulties of the UK fishing industry. The consumer would also be likely to see some increases in fish prices. However, the impacts within the United Kingdom would be unlikely to be major since the UK fishing industry now employs relatively few people and fish has formed a declining component of the national diet.

Materials

A number of commercially important materials that are widely used in construction and other applications are adversely affected by solar ultraviolet radiation. The materials affected include synthetic polymers such as plastics, naturally occurring biopolymers, particularly rubber and timber products and protective materials such as paints and varnishes. Any increase in solar ultraviolet radiation as a results of stratospheric ozone depletion would be likely to increase the rate of photodegradation of such materials (Andrady *et al*., 1995). One of the earliest effects of photodamage is likely to be discolouration which may set an effective limit to serviceability for applications where cosmetic appearance is important. Further exposure can lead to a loss of mechanical integrity which can set a limit to service lifetimes even where appearance is not critical. Decreases in the useful lifetime of materials such as plastics and timber products used in construction will add to costs since such materials will have to be replaced more frequently. Costs of protective maintenance such as painting are also likely to rise. It is difficult to give precise estimates of the magnitude of such additional costs. Although action spectra for photodamage are available for some materials the available information is often inadequate as a basis for estimating increases in damage. Another area of uncertainty relates to the development of new products that might be less subject to photodamage than those that are currently used. Many existing products are serviceable in the outdoor environment only because of the incorporation of light-stabilizers in polymer formulations. Some of the effects of enhanced ultraviolet radiation exposure might be mitigated by greater use of light stabilizers. However, these materials already constitute an important part of the price of many products and any increase in their use would be likely to add to costs Another approach would be substitution of the affected materials by other materials that are more photostable, but again this might add to costs.

10.5 Conclusions

Increases in UV-related diseases suggest that population exposure to solar ultraviolet radiation has increased. Changes in levels of ultraviolet radiation in the environment have so far played little part in this increase. The main factors have probably been changes in behaviour including the growth in foreign holidays, increased participation in outdoor leisure activities, changes in clothing and the continuing popularity of sunbathing and a sun-tanned appearance. However, higher levels of UV in the future could increase some health risks. It is important to stress that there is potential for any human health effects of increases in levels of UV radiation in the environment to be mitigated by changes towards more cautious behaviour. However, there are some important obstacles to such changes.

Socio-economic implications of possible health impacts include costs of treatment, costs of programmes of prevention, the value of lives that are lost and the effects on the tourism and leisure industry of any change in behaviour resulting from perceptions of increased risks. There might also be economic implications arising from the effects of enhanced UV on agriculture, forests, fishing and on materials. However, precise estimates of such impacts are not yet possible. Much of this imprecision arises from uncertainties about future levels of UV and a lack of basic knowledge of the impact of UV on important physical and biological systems. Furthermore, any impacts of such changes on human welfare are likely to be strongly affected by social and economic factors.

References

Acheson ED (1986). From the Chief Medical Officer: Mortality from cutaneous malignant melanoma. *Health Trends* 18, 73.

Adami J, Frisch M, Yuen J, Glimelius B and Melbye M. (1995) Evidence of an association between non-Hodgkin's lymphoma and skin cancer. *British Medical Journal* 310, 1491-1495.

Anderson DE and Badzioch C (1991) Association between solar radiation and squamous cell carcinoma in cattle. *American Journal of Veterinary Research* 52, 784-788.

Andrady AL, Torikai A and Fueki K (1989) Photodegradation of rigid PVC formulations, I: Wavelength sensitivity to light-induced yellowing by monochromatic light. *Journal of Applied Polymer Science* 37 (4), 935-946.

Andrady AL (1991) Spectral sensitivity of polycarbonate to light-induced yellowing. *Journal of Applied Polymer Science* 42 (7), 2105-2107.

Andrady AL, Amin MB, Hamid SH, Hu X and Torikai A (1994). Effects of increased solar radiation on materials. In: *UNEP Environmental Effects on Ozone Depletion: 1994 Assessment*. Chapter 7, United Nations Environment Programme, Nairobi, Kenya, November 1994 (ISBN 92-807-1457-0).

Andrady AL, Amin MB, Hamid SH, Hu X and Tokai A (1995) Effects of increased solar ultraviolet radiation on materials. *Ambio* 24, 191-6.

Armstrong BK (1994). Stratospheric ozone and health. *International Journal of Epidemiology* 23, 873-885.

Armstrong BK and Kricker A (1994) Cutaneous Melanoma. *Cancer Surveys* 19, 219-240.

Armstrong BK and Kricker A (1995) Skin Cancer. *Dermatoepidemiology* 13, 583-594.

Atkinson R (1990) Gas phase tropospheric chemistry of organic compounds: a review. *Atmospheric Environment* 24A, 1-42.

Austin JN, Butchart and Shine KP (1992) Possibility of an Arctic ozone hole in a doubled CO_2 climate. *Nature* 360, 221-225.

Austin J, Barwell BR, Cox SJ, Hughes PA, Knight JR, Ross G, Sinclair P and Webb AR (1994). The diagnosis and forecast of clear sky ultraviolet levels at the Earth's surface. *Meterological Applications* 1, 321-336.

Autier P, Dore JF, Schiffler E *et al.* (1995) Melanoma and use of sun screens: an EORTC case-control study in Germany, Belgium and France. *International Journal of Cancer* 6, 1749-55.

Barnes WB, Flint SD and Caldwell MM (1995) Early season effects of supplemented solar UV-B radiation on seedling emergence, canopy structure, simulated stand photosynthesis and competition for light. *Global Climate Change* 1, 43-54.

Behrenfeld MJ, Lean DRS and Lee II H (1995) Ultraviolet-B radiation effects on inorganic nitrogen uptake by natural assemblages of oceanic plankton. *Journal of Phycology* 31, 25-36.

Bekki S, Law KS and Pyle JA (1994) Effect of ozone depletion on atmospheric CH_4 and CO concentrations. *Nature* 371, 595-597.

Blaustein AR, Hoffman PD, Hokit DG, Kiesecker JM, Walls SC and Hays JB (1994). UV repair and resistance to solar UV-B in amphibian eggs: A link to population declines? *Proceedings of the National Academy of Science USA* 91, 1791-1795.

Blumthaler M, Ambach W, Silbernagl R and Staehelin J (1994), Erythemal UV-B irradiance (Robertson-Berger sunburn metre data) under ozone deficiencies in Winter/Spring. *Photochemistry and Photobiology* 59, 657-659.

Blumthaler M, Webb AR, Seckmeyer G, Bais AF, Huber M and Mayer B (1994). Simultaneous spectroradiometry: a study of solar UV irradiance at two altitudes. *Geophysical Research Letters* 21, 2805-2808.

Bootsma D and Hoeijmakers JHJ (1993). Engagement with transcription. *Nature* 363, 114-115.

Bornman JF, Björn LO and Akerlund H-E (1984). Action spectrum for inhibition by ultraviolet radiation of photosystem II activity in spinach thylakoids. *Photobiochemistry and Photobiophysics* 8, 305-313.

Bornman JF and Teramura AH (1993). Effects of ultraviolet-B radiation on terrestrial plants. pp. 427-471. In: *Environmental UV Photobiology*. Young AR, Bjorn LO, Moan J and Nultsch W (Eds.) New York: Plenum Press.

Bothwell ML, Sherbot DM and Pollock CM (1994). Ecosystem response to solar ultraviolet-B radiation: influence of trophic-level interaction. *Science* 265, 97-100.

Bothwell ML, Karentz D and Carpenter EJ (1995). No UV-B effect? *Nature* 374, 601.

British Medical Association (1987). *Living with risk*. London: British Medical Association.

Bruhl C and Crutzen PJ (1989). On the disproportionate role of tropospheric ozone as a filter against solar UV-B radiation, *Geophysical Research Letters* 16, 703-706.

Bullock AM (1982). The pathological effects of ultraviolet radiation on the epidermis of teleost fish with reference to solar radiation effect in higher animals. *Proceedings of the Royal Society of Edinburgh* 81B, 199-210.

Bullock AM (1988). Solar ultraviolet radiation: a potential environmental hazard in the cultivation of farmed fish. In: *Recent Advances in Aquaculture*, Vol 3, Muir J and Roberts RJ (Eds.) Timber Press, Portland. p.420.

Burra AGJ, van Hannen EJ, Roza L, Veldhuis MJW and Gieskes WWC (1995). Monitoring ultraviolet-B-induced DNA damage in individual diatom cells by immunofluorescent thymine dimer formation. *Journal of Phycology* 31, 314-321.

Burrows WR, Vallee M, Wardle DI, Kerr JB, Wilson LJ and Tarasick DW (1994). The Canadian operational procedure for forecasting total ozone and UV radiation. *Meteorological Applications* 1, 247-265.

Caldwell MM (1971). Solar UV irradiation and the growth and development if higher plants. pp.131-177 in *Photophysiology*, Giese AC (Ed) Academic Press, New York.

Caldwell MM, Camp LB, Warner CW and Flint SD (1986). Action spectra and their key role in assessing biological consequences of solar UV-B radiation change. In: *Stratospheric ozone reduction, solar ultraviolet radiation and plant life: Workshop on the impact of solar ultraviolet radiation upon terrestrial ecosystems: 1. Agricultural crops,* pp 87-111. Worrest RC and Caldwell MM (Eds), Springer-Verlag, New York.

Caldwell MM, Teramura AH, Tevini M, Bornman JF, Bjorn LO and Kulandaivelu G (1989). Terrestrial plants. pp. 25-37. In: *Environmental effects of ozone depletion*, 1989 Assessment, UNEP Environmental effects Panel Report, Nairobi, Kenya.

Caldwell MM, Teramura AH, Tevini M, Bornman JF, Bjorn LO and Kulandaivelu G (1995). Effects of increased solar ultraviolet radiation of terrestrial plants. *Ambio* 24, 166-173.

Calkins J (Ed.) (1982). *The Role of Solar Ultraviolet Radiation in Marine Ecosystems*. NATO Conference Series. Plenum Press, New York, London.

Cartwright R, McNally R and Staines A (1994). The increasing incidence of non-Hodgkin's lymphoma (NHL): the possible role of sunlight. *Leuk Lymph* 14, 387-394.

Chan GL, Peak MJ, Peak JG and Haseltine WA (1986). Action spectrum for the formation of endonuclease-sensitive sites and (6-4) photoproducts induced in a DNA fragment by ultraviolet radiation. *International Journal of Radiation Biology* 50, 641-648.

Cofino WP, Smedes F, de Jong SA, Abarnou A, Boon JP, Davies IM, Klungsoyr J, Wilhelmsen S, Law, RJ Whinnet JA, Schmidt D and Wilson S (1992). The chemistry programme. *Mar. Ecol. Prog. Ser.* 91, 47-56.

Cooper KD, Oberhelman L, Hamilton TA, Baadsgaard O, Terhune M, Levee G, Anderson T and Koren H (1992). UV exposure reduces immunization rates and promotes tolerance to epicutaneous antigens in humans: Relationship to dose, CD1a-DR+ epidermal macrophage induction, and Langerhans cell depletion. *Proceedings of the National Academy of Science USA* 89, 8497-8501.

Corlett JE, Stephen J, Jones HJ, Woodfin R, Mepsted R and Paul ND (1997). Assessing the impact of UV-B on the growth and yield of field crops, In Plant and UV-B: Responses to Environmental Change, Lumsden PJ (Ed.), Cambridge University Press, in press.

Cullen JJ, Neale PJ and Lesser MP (1992). Biological weighting functions for the inhibition of phytoplankton photosynthesis by ultraviolet radiation. *Science* 258, 646-650.

Davis A, Deane GHW and Diffey BL (1976). *Nature* (London) 261, 169.

Davis A and D Sims (1983). Weathering of Polymers. Applied Science Publishers, London and New York.

Davis A and Gardiner D (1987). *Polymer Testing* 7, 345.

References

De Fabo EC and Noonan FP (1983). Mechanism of immune suppression by ultraviolet radiation *in vivo*. I: Evidence for the existence of a unique photoreceptor in skin and its role in photoimmunology. *Journal of Experimental Medicine* 158, 84-98.

De Fabo EC, Reilly DC and Noonan FP (1990). Biologically effective doses of sunlight for immunosuppression at various latitudes and their relationship to changes in stratospheric ozone. *Photochemistry and Photobiology* 52, 811-817.

De Gruijl FR, Sterenborg HJCM, Forbes PD, Davies RE, Cole C, Kelfkens G, van Weelden H, Slaper H and van der Leun JC (1993). Wavelength dependence of skin cancer induction by ultraviolet irradiation of albino hairless mice. *Cancer Research* 53, 53-60.

De Gruijl FR and van der Leun JC (1994). Estimate of the wavelength dependency of ultraviolet carcinogenesis in humans and its relevance to the risk assessment of a stratospheric ozone depletion. *Health Phys*. 67 (4), 319-325.

De Leeuw FAAM and Von Rheineck Leysius (1991) Sensitivity of oxidant concentrations on changes in UV radiation and temperature. *Atmospheric Environment* 25A, 1025-1032.

Demerjian KL, Kerr JA and Calvert JG (1974) The mechanism of photochemical smog formation. *Advances in Environmental Science and Technology* 4, 1-262.

Diffey BL (1987). Analysis of the risk of skin cancer from sunlight and solaria in subjects living in northern Europe. *Photodermatology* 4,118-26.

Diffey BL (1992). Stratospheric ozone depletion and the risk of non-melanoma skin cancer in a British population. *Phys Med Biol* 37, 2267-2279.

Diffey BL, Gibson CJ, Haylock R and McKinlay AF (1996). Outdoor ultraviolet exposure of children and adolescents. *British Journal of Dermatol*ogy, in press.

Dlugokencky EJ, Masarie KA, Lang PM, Tans PP, Steele LP and Nisbet EG (1994) A dramatic decrease in the growth rate of atmospheric methane in the northern hemisphere during 1992. *Geophysical Research Letters* 21, 45-48, Correction p.507.

Doniger J, Jacobson ED, Krell K and DiPaolo JA (1981). Ultraviolet light action spectra for neoplastic transformation and lethality of Syrian hamster embryo cells correlate with spectrum for pyrimidine dimer formation in cellular DNA. *Proceedings of the National Academy of Sciences of the USA* 78, 2378-2382.

Döhler G (1984). Effects of UV-B radiation on the marine diatoms *Lauderia annulata* and *Thalassiosira rotula* grown in different salinities. *Marine Biology* 83, 247-253.

Dorn CR, Taylor DON and Schneider R (1971) Sunlight exposure and risk of developing cutaneous and oral squamous cell carcinomas in white cats. *Journal of National Cancer Institute* 46, 1073-1078.

Driscoll CMH and Rawlinson AI (1995). Weekly Solar UVR Bulletins, *National Radiological Protection Board Bulletin* 166, 8-11.

Driscoll CMH (1996). *National Radiological Protection Board Bulletin* 176, p.4.

D'Surney SJ, Tschaplinski TJ, Edwards NT and Shugart LR (1993) Biological responses of two soybean cultivars exposed to enhanced UVB radiation. *Environmental and Experimental Botany* 33, 347-336.

El-Ghorr AA and Norval MA (1995). Monoclonal antibody to cis-urocanic acid prevents the ultraviolet radiation-induced changes in Langerhans cells and delayed hypersensitivity in mice, although not preventing dendritic cell accumulation in lymph nodes draining site of irradiation and contact hypersensitivity responses. *Journal of Investigative Dermatology* 105, 264-268.

Ellison MJ and Childs JD (1981). Pyrimidine dimers induced in *Escherichia coli* DNA by ultraviolet radiation present in sunlight. *Photochemistry and Photobiolgy* 34, 465-469.

Elmets CA, LeVine MJ and Bickers DR (1985). Action spectrum studies for induction of immunologic unresponsiveness to dinitrofluorobenzene following in vivo low dose ultraviolet radiation. *Photochemistry and Photobiol*ogy 42, 391-397.

Elwood JM (1992). Melanoma and sun exposure: contrast between intermittent and chronic exposure. *World Journal of Surgery* 16, 157-65.

Enninga IC, Groenendijk RTL, Filon AR, vanZeeland AA and Simons JWIM (1986). The wavelength dependence of UV-induced pyrimidine dimer formation, cell killing and mutation induction in human diploid skin fibroblasts. *Carcinogenesis* 7, 1829-1836.

Farmer SFG (1993). The annual rate decrease of total column ozone observed at Lerwick and Bracknell/Camborne compared with TOMS satellite data, *Observation Provision Branch Memorandum* No. 129, Meteorological Office, Bracknell,.

Fiscus EL, Miller JE and Booker FL (1992) Is UV-B a hazard to soybean photosynthesis and yield? Results of an ozone-UV-B interaction study and model predictions. pp. 135-147 In: *Stratospheric ozone depletion/UV-B radiation in the biosphere*. Biggs RH & Joyner MEB (Eds.) NATO-ASI Series, Springer-Vorlage, Berlin.

Fiscus EL and Booker FL (1995) Is increased UV-B a threat to crop photosynthesis and photosynthesis? *Photosynthesis Research* 43, 81-92.

Friedberg EC, Walker GC and Siede W (1995). *DNA Repair and Mutagenesis*, Washington, DC, American Society for Microbiology.

Fuglestvedt JS, Jonson JE and Isaksen ISA (1994) Effects of reductions in stratospheric ozone on tropospheric chemistry through changes in photolysis rates. *Tellus* 46B,172-192.

Gallagher RP, Hill GB, Bajdik CD, Fincham S, Coldman AJ, McLean DI and Threlfall WJ (1995a). Sunlight Exposure, Pigmentation Factors, and Risk of Nonmelanocytic Skin Cancer: II Squamous Cell Carcinoma. *Archives of Dermatology* 131, 164-169.

Gallagher RP, Hill GB, Bajdik CD, Fincham S, Coldman AJ, McLean DI and Threlfall WJ (1995b). Sunlight Exposure, Pigmentary Factors, and Risk of Nonmelanocytic Skin Cancer: I Basal Cell Carcinoma. *Archives of Dermatology* 131, 157-163.

Garcia-Pichel F and Castenholz RW (1991). Characterisation and biological implications of scytonemin and cyanobacterial sheath pigment. *Journal of Phycology* 27, 395-409.

Garcia-Pichel F (1994). A model for internal self-shading in planktonic organisms and its implication for the usefulness of ultraviolet sunscreens. *Limnol. Oceanogr.* 39, 1704-1717.

Gehrke C, Johnason U, Callaghan TV, Chadwick D and Robinson CH (1995) The impact of enhanced ultraviolet-B radiation on litter quality and decomposition processes in *Vaccinium* leaves from the subarctic. *Oikos* 71, 213-222.

Gery M (1989) Tropospheric air quality. Chapter 5. Environmental Effects Panel Report. United Nations Environment Programme, Niarobi, Kenya.

Gieskes WC and Kraay GW (1990). Transmission of ultraviolet light in the Weddell Sea. Report on the first measurements made in the Antarctic. *Biomass Newsletter* 12, 12-14.

Glass AG and Hoover RN (1989). The emerging epidemic of melanoma and squamous cell skin cancer. *Journal of the American Medical Association* 262, 2097-2100.

Goettsch W, Garssen J, de Gruijl FR and van Loveren H (1994) Effects of UVB on the resistance against disease. *Toxicology Letters* 72, 359-363.

Goettsch W, Garssen J, Hurks HMW, Slob W, Dortant P, de Gruijl FR and van Loveren H (1995). Susceptibility of different species for UVB radiation with respect to the immune system. *Photochemistry and photobiology* 61S, 54 (abstract).

Gold WG and Caldwell MM (1983) The effects of ultraviolet-B radiation on plant competition in terrestrial ecosystems. *Physiologia Plantarum* 58, 435-444.

Grabbe S and Granstein RD (1995). Photoimmunology of UV-induced skin cancer. In: *Photoimmunology*. J Krutman and CA Elmets (Eds.), 99-118. Blackwell Science, Oxford.

Green AES, Sawada T and Shettle EP (1974). The middle ultraviolet reaching the ground. *Photochemistry and Photobiology* 19, 251-259.

De Gruijl FR and Forbes PD (1995) UV-induced skin cancer in a hairless mouse model. *Bioessays* 17, 651-660.

Haagen-Smit AJ, Bradley CE and Fox MM (1953) Ozone formation in photochemical oxidation of organic substances. *Industrial Engineering Chemistry* 45, 2086-2089.

Häder DP (1993). Risks of enhanced solar ultraviolet radiation for aquatic ecosystems. *Progr. Phycol. Res.* 9, 1-45.

References

Häder D-P, Worrest RC and Kumar HD (1995). Effects of increased solar ultraviolet radiation on aquatic ecosystems. *Ambio* 24, 174-180.

Häder DP, Herrmann H and Santos R (1996). Effect of solar radiation and solar radiation deprived of UV-B and total UV on photosynthetic oxygen production and pulse amplitude modulated fluorescence in the brown alga *Padina pavonia FEMS Microbial Ecology* 19, 53-61.

Health Council of the Netherlands (1994). Risks of UV Radiation Committee. UV radiation from sunlight. The Hague: Health Council of the Netherlands, Publication no. 1994/05E

Højerslev NK (1988). Natural occurrences and optical effects of Gelbstoff. Københavns Universitet Geofysisk Institut. Copenhagen Report No.50, 30pp.

Honeyman J, Koren HS and West S (1995) Ocular and dermatologic health effects of ultraviolet radiation exposure from the ozone hole in southern Chile. *American Journal of Public Health* 85, 546-550.

Hurks MHH, Out-Luiting C, Vermeer BJ, Claas FHJ and Mommaas H (1995). The action spectra for UV-induced suppression of MLR and MECLR show that immunosuppression is mediated by DNA damage. *Photochemistry and Photobiology* 62, 449-453.

IARC Monogrpahs on the evaluation of carcinogenic risks to humans: Volume 55, pp 316, Solar and Ultraviolet Radiation, IARC, Lyon, 1992.

Jeevan A, Brown E and Kripke ML (1995). UV and infectious diseases. In: *Photoimmunology*. Eds. J Krutman and CA Elmets, pp153-163. Blackwell Science, Oxford.

Joint I and Pomeroy A (1993). Phytoplankton biomass and production in the southern North Sea. *Mar. Ecol. Progr. Ser.* 99, 169-182.

Kacser M and Burns JA (1979). Molecular democracy: who shares the controls. *Biochemistry Society Transactions* 7, 1149-1160.

Karentz D (1994). Ultraviolet tolerance mechanisms in Antarctic marine organisms. In: *Ultraviolet Radiation in Antarctica: Measurements and Biological Effects*. Weiter CS & Penhale PA (Eds.), pp. 93-110. AGU, Washington, DC.

Kessler E and Björn LO (Eds.) (1995). Further Research into UV-B Radiation. *Ambio* 24, 3 Special Issue.

Khalil MAK and Rasmussen RA (1994). Global decrease in atmospheric carbon monoxide concentration. *Nature* 370, 639-641.

Kiesecker JM and Blaustein AR (1995). Synergism between UV-B radiation and a pathogen magnifies amphibian embryo mortality in nature. *Proceedings of the National Academy of Science USA* 92, 11049-11052.

Kimura T and Doi K (1994). Responses of the skin over the dorsum to sunlight in hairless descendants of Mexican hairless dogs. *American Journal of Veterinary Research* 55, 199-203.

Kirk JTO (1994a). *Light and photosynthesis in aquatic ecosystems*. Second Edition. Cambridge University Press.

Kirk JTO (1994b). Optics of UV-B in natural waters. *Arch. Hydrobiol. Beih.* 43, 1-16.

Klein KL and Linnemann CC (1986). Induction of recurrent genital herpes simplex virus type 2 infection by ultraviolet light. *The Lancet* I, 796-797.

Kligman LH and Sayre RM (1991). An action spectrum for ultraviolet induced elastosis in hairless mice: Quantification of elastosis by image analysis. *Photochemistry and Photobiology* 53, 237-242.

Klironomos JN and Allen MF (1995). UV-B mediated changes on below-ground communities associated with the roots of Acer saccharum. *Functional Ecology* 9, 923-930.

Kramer KJM (1990). Effects of increased solar UV-B radiation on coastal marine ecosystems. In: *Expected Effects of Climate Changes on Marine Ecosystems.* Beukema *et al.* (Eds). 195-210. Kluwer Academic Publishers.

Kress S, Sutter C, Strickland PT, Mukhtar H, Schweizer J and Schwarz M (1992). Carcinogen-specific mutational pattern in the P53 gene in ultraviolet B radiation-induced squamous carcinomas of mouse skin. *Cancer Research* 52, 6400-6403.

Kricker A, Armstrong BK and English DR (1994). Sun exposure and non-melanocytic skin cancer. *Cancer Causes and Control* 5, 367-392.

Kricker A, Armstrong BK, English DR and Heenan PJA (1995) Dose-Response curve for sun exposure and basal cell carcinoma. *International Journal of Cancer* 60, 482-488.

Kricker A, Armstrong BK, English DR and Heenan PJ (1995). Does intermittent sun exposure cause basal cell carcinoma? A case-control study in Western Australia. *International Journal of Cancer* 60, 489-494.

Kripke ML, Cox PA, Alas LG and Yarosh DB (1992). Pyrimidine dimers in DNA initiate systemic immunosuppression in UV-irradiated mice. *Proceedings of the National Academy of Science USA* 89, 7516-7520.

Krupa SV and Kickert RN (1989) The greenhouse effect- impacts of ultraviolet-B (UV-B) radiation, carbon dioxide (CO_2) and ozone (O_3) on vegetation. *Environmental Pollution* 61, 263-393.

Laane RWPM and Kramer KJM (1990). Natural fluorescence in the North Sea and its major estuaries. *Netherlands Journal of Sea Research* 26, 1-9.

Landrum PF, Giesy JP, Oris JT and Allred PM (1986). Photoinduced toxicity of polycyclic aromatic hydrocarbons to aquatic organisms. In: *Oil in Fresh Water: Chemistry, Biology, Technology*. Vandermuelen L & Hrudy (Eds.) Pergamon Press, NY pp, 324-328.

Larkum AWD and Wood WF (1993). The effect of UV-B radiation on photosynthesis and respiration of phytoplankton, benthic macroalgae and seagrasses. *Photosynthesis Research* 36, 17-23.

Leighton PA (1961). Photochemistry of air pollution. Academic Press, New York.

Little EE and Fabacher DL (1994). Comparative sensitivity of rainbow two and two threatened salmonids, Apache trout and Lahontan cutthroat trout, to ultraviolet radiation. *Ergebn. Limnol.* 43, 217-226.

Liu SC and Trainer M (1988). Response of the tropospheric ozone and odd hydrogen radicals to column ozone change. *Journal of Atmospheric Chemistry* 6, 221-233.

Long LE, Saylor LS and Soule ME (1995). A pH/UV-B Synergism in Amphibians. *Conservation Biology* 9, 1301-1303.

Lubin D and Jensen EH (1995). Effects of clouds and stratospheric ozone depletion on ultraviolet radiation trends. *Nature* 377, 710-713.

McCormick MP, Veiga RE and Chu WP (1992). Stratospheric ozone profile and total ozone trends derived from the SAGE I and SAGE II data, *Geophysical Research Letters* 19, 269-272.

McGregor JM and Proby CM (1995). Skin cancer in transplant recipients. *The Lancet* 346, 964-965.

McKinlay AF and Diffey BL (1987). A reference action spectrum for ultra-violet induced erythema in human skin CIE J (6), 1, 17-22. In: *Human exposure to ultraviolet radiation: Risks and regulations*, Passchier WF and Bosnjakovic BFM (Eds.), Elsevier Science Publishers BV (Biomedical Division), Amsterdam, 83-87.

McLeod AR (1996). Outdoor supplementation systems for studies of the effects of increased UV-B radiation. pp. In: *UV-B and Biosphere*, Rozema J, Gieskes W, van de Geijn SC, Nolan C and de Boois H (Eds), Kluwer Academic Publishers, Dordrecht.

McMinn A, Heijnis H and Hodgson D (1994). Minimal effects of UVB radiation on Antarctic diatoms over the past 20 years. *Nature* 370, 547-549.

Madewell BR and Theilen GH (1987) In: *Veterinary Cancer Medicine*. Theilen GH and Madewell BR (Eds.), Leas & Febiger, Philadelphia.

Madronich S, McKenzie RL, Caldwell MM, and Björn LO (1995). Changes in Ultraviolet Radiation Reaching the Earth's Surface. *Ambio* 24 (3), 143-152.

Magnus K (1991). The Nordic profile of skin cancer incidence: A comparative epidemiological study of the three main types of skin cancer. *Interna tional Journal of Cancer* 47,12-19.

Manning WJ and Teidemann AV (1995) Climate change: potential effects of increased atmospheric carbon dioxide (CO_2), ozone (O_3) and ultraviolet-B radiation (UV-B) on plant diseases. *Environmental Pollution* 88, 219-245.

Marchant HJ, Davidson AT and Kell GJ (1991). UV-B protecting compounds in the marine alga *Phaeocystis pouchetii* from Antarctica. *Marine Biology* 109, 391-395.

Matsunaga T, Hieda K and Nikaido O (1991). Wavelength dependent formation of thymine dimers and (6-4) photoproducts in DNA by monochro matic ultraviolet light ranging from 150 to 365 nm. *Photochemistry and Photobiology* 54, 403-410.

References

Mepsted R, Stephen J, Corlett JE, Nogues S, Ayres PG, Jones HG, Baker NR and Paul ND (1995). Effects of enhanced UV-B radiation on pea (*Pisum sativum* L.) grown under field conditions in the United Kingdom. *Global Climate Change*, submitted.

Mitchell DL, Pfeifer GP, Taylor JS, Zdienicka MZ and Nikaido O (1993). Biological role of (6-4) photoproducts and cyclobutane dimers. pp 337-344. In: *Frontiers of Photobiology*, Shima A, Ichihashi M, Fujiwara Y, Takebe H. (Eds.), Amsterdam, Elsevier.

National Radiological Protection Board (1995). *Health Effects from Ultraviolet Radiation*. Report of an Advisory Group on Non-ionising Radiation, Documents of the NRPB, Vol 6, No 2, p155.

Newstead JL and Giesy JP (1987). Predictive models for photoinduced acute toxicity of polyaromatic hydrocarbons to *Daphnia magna* Strauss (Cladocera, Crusatacea). *Environ. Toxicol. Chem.* 6, 445-461.

Nordic Council of Ministers (1991) Photochemical oxidants in the atmosphere. Nord 7, Copenhagen, Denmark.

Norval M, El-Ghorr A, Garssen J and van Loveren H (1995). The effects of ultraviolet radiation on viral infections. *British Journal of Dermatology* 130, 693-700.

Novelli PC, Masarie K, Tans PP and Lang P (1994) Recent changes in atmospheric carbon monoxide. *Science* 263, 1587-1590.

NSQSR (1993). North Sea Quality Status Report. Oslo and Paris Commissions, London.

Palenik B, Price NM and Morel FMM (1991). Potential effect of UV-B on the chemical environment of marine organisms: a review. *Environmental Pollution* 70, 117-130.

Pathak MA and Fanselow DL (1983). Photobiology of melanin pigmentation: dose/response of skin to sunlight and its contents, *J. Am. Acad. Dermatol.* 9, 724-733.

Paul ND, Rasanayagam MS, Moody SA, Hatcher PE and Ayres PG (1996) The role of interactions between trophic levels in determining the effects of UV-B on terrestrial ecosystems. *Vegetatio*, submitted

Peak MJ, Peak JG, Moehring MP and Webb RB (1984). Ultraviolet action spectrum for DNA dimer induction, lethality, and mutagenesis in *Escherichia coli* with emphasis on the UVB region. *Photochemistry and Photobiology* 40, 613-620.

Pearce DW, Bann C and Georgiou S (1992). *Social cost of fuel cycles*. London: HMSO.

Philipov JP (1992) Changes in some biochemical indicators of bone turnover after ultraviolet irradiation of dairy cows. *Research in Veterinary Science* 53, 397-398.

Pingree RD and Griffiths DK (1978). Tidal fronts on the shelf seas around the British Isles. *Journal of Geophysical Research* 83 (C9), 4615-4622.

Pitts DG, Cullen AP and Hacker PD (1977). Ocular effects of ultraviolet radiation from 295 to 365 nm. *Invest. Ophthalmol. Visual Sci.* 16, 932-939.

PORG (1993). *Ozone in the United Kingdom 1993*, Photochemical oxidants Review Group, Department of the Environment, London.

Rabek JF (1990). *Photostabilization of Polymers*. Elsevier Applied Science, London and New York.

Rao MV and Ormrod DP (1995). Ozone exposure decrease UV-B sensitivity in a UV-B sensitive flavonoid mutant of Arabidopsis. *Photochemistry & Photobiology* 61, 71-78.

Raven JA (1991). Responses of aquatic photosynthetic organisms to increased solar UV-B. *J. Photochem. and Photobiol. B: Biol.* 9, 239-244.

Rooney JF, Bryson Y, Mannix ML, Dillon M, Wohlenberg CR, Banks S, Wallington CJ, Notkins AL and Straus SE (1991). Prevention of ultraviolet-light-induce herpes labialis by sunscreen. *The Lancet* 338, 1419-1422.

Rosenstein BS and Mitchell DL (1991). The repair of DNA damages induced in normal human skin fibroblasts exposed to simulated sunlight. *Radiation Research* 126, 338-342.

Rozema J, Lensson GM, Tosserams M and Visser AJ (1996). Effects of UV-B on terrestrial plants. Interactions with atmospheric CO_2 enrichment. In: *UV-B and Biosphere*. Rozema J, Gieskes W, van de Geijn SC, Nolan C & de Boois H. (Eds) Kluwer Academic Publishers, Dordrecht.

Scatter DH, Lavach JD, Severin GA and Young S (1977) Überreiter's Syndrome (chronic superficial keratitis) in dogs in the Rocky Mountain area -a study of 463 cases. *Journal of Small Animal Practice* 18, 757-772.

Schein OD, Vicencio C, Munoz B, Gelatt KN, Duncan DD, Nethercott J,

Schindler DW, Curtis PJ, Parker BP and Stainton MP (1996). Consequences of climate warming and lake acidification for UV-B penetration in North American boreal lakes. *Nature* 379, 705-708.

Scotto J, Picher H and Lee JAH (1991). Indications of future decreasing trends in skin-melanoma mortality among whites in the United States. *International Journal of Cancer* 49, 490-7.

Setlow RB (1966). Cyclobutane-type pyrimidine dimers in polynucleotides. *Science* 153, 379.

Setlow RB (1974). The wavelengths in sunlight effective in producing skin cancer: A theoretical analysis. *Proceedings from the National Academy of Science USA* 71 (9), 3363-3366.

Shearer JK, Tesopgoni T and Giggs EPJ (1992). In: *Bovine Medicine*, Chapter 25. Andrews AH, Blowey RW, Boyd H & Eddy RG (Eds.). Blackwell Scientific Publications, Oxford.

Smith RC, Prézelin BB, Baker KS, Bidigare RR, Boucher NP, Coley T, Karentz D, MacInture S, Matlick MA, Menzies D, Ondrusek M, Wan Z and Waters KJ (1992). Ozone depletion: ultraviolet radiation and phytoplankton biology in Antarctic Waters. *Science* 255, 952-959.

SORG (1993). Stratospheric Ozone 1993. Fifth report of the United Kingdom Stratospheric Ozone Review Group. HMSO, London. ISBN 0 11 752861 7.

SORG (1996) Stratospheric Ozone 1996. Sixth report of the United Kingdom Stratospheric Ozone Review Group. Department of the Environment reference number 96DPL0020.

Spruance SL (1985). Pathogenesis of herpes simplex labialis: Experimental induction of lesions with UV light. *Journal of Clinical Microbiology* 22, 366-368.

Streilein JW (1995). Immunogenetics of sunlight-induced skin cancer. In: *Photoimmunology*. Krutman J and Elmets CA (Eds.), pp. 128-152, Blackwell Science, Oxford.

Sullivan JH and Teramura AH (1990). Effects of UV-B radiation on soybean yield and seed quality: a six year field study. *Physiologia Plantarum* 80, 5-11.

Suozzo RD, Balachandran NK and Prather MJ (1990). Climate change and the middle atmosphere. Part I: The doubled CO_2 climate, *Journal of Atmospheric Science* 47, 475-494.

Tang X and Madronich S (1994). Effects of increased solar ultraviolet radiation on tropospheric composition and air quality. Chapter 6. Environ mental Effects of Ozone Depletion: 1994 Assessment. United Nations Environment Programme, Nairobi, Kenya.

Taylor HR (1995). Ocular effects of UV-B exposure. *Documenta Ophthalmologica* 88, 285-293.

Teramura AH, Tevini M, Bornman JF, Caldwell MM, Kulandaivelu G and Bjorn LO (1991). Terrestrial plants. In: *Environmental effects of ozone depletion*, pp. 25-32, 1991 Update, UNEP Environmental effects Panel Report, Nairobi, Kenya.

Thimijan RW, Carns HR and Campbell LE (1978). *Radiation sources and related environmental control for biological and climatic effects of UV research* (BACER). Final report EPA-IAG-D6-0168. Environmental Protection Agency, Washington, DC.

Thompson AM, Huntley MA and Stewart RW (1991). Perturbation to tropospheric oxidants, 1985-2035. Calculations of ozone and OH in chemically-coherent regions. *Atmospheric Environment* 25A, 1837-1850.

Tyrrell RM and Pidoux M (1988). Correlation between endogenous glutathione content and sensitivity of cultured human skin cells to radiation at defined wavelengths in the solar ultraviolet range. *Photochemistry and Photobiology* 47, 404-412.

Ullrich SE (1995). The role of epidermal cytokines in the generation of cutaneous immune reactions and ultraviolet radiation-induced immune suppression. *Photochem and Photobiol* 62, 389-401.

UNEP (1989). Environmental Effects Panel Report. United Nations Environment Programme. Nairobi, Kenya.

UNEP (1991). Environmental Effects of Stratospheric Ozone Depletion -1991 Update. United Nations Environment Programme. Nairobi, Kenya.

References

UNEP (1994). Environmental Effects on Ozone Depletion: 1994 Assessment. United Nations Environment Programme, Nairobi, Kenya, November 1994 (ISBN 92-807-1457-0).

Webb AR (1991). Solar ultraviolet radiation in southeast England: The case for spectral measurements. *Photochemistry and Photobiology* 54, 789-794.

Westerdahl J, Olsson H, Måsbäck A, Ingvar C and Jonsson N (1995). Is the use of sunscreens a risk factor for malignant melanoma. *Melanoma Research* 5, 59-65.

Wheeler CE (1975). Pathogenesis of recurrent herpes simplex infections. *Journal of Investigative Dermatology* 65, 341-346.

World Health Organization (1994). *Ultraviolet radiation. Environmental Health Criteria 160*, Geneva: WHO, Table 1 Standardised proportional registration ratios for malignant melanoma of the skin for males aged 15-74 in England and Wales, 1974-84.

World Meteorological Organization (1995). Scientific Assessment of Ozone Depletion: 1994. WMO Global Ozone Research and Monitoring Project, Report No. 37. World Meteorological Organization, Geneva. ISBN 92 807 1449 X.

Worrest RC (1982). Review of literature concerning the impact of UV-B radiation upon marine organisms. In: *The Role of Solar Ultraviolet Radiation in Marine Ecosystems*. Calkins J (Ed.) NATO Conference Series. Plenum Press, New York & London.

Young RW (1994). The family of sunlight-related eye diseases. *Optometry and Visual Science* 71, 125-144.

Zaneveld JRV (1975). Penetration of ultraviolet radiation into natural waters. In: *Impacts of climatic change on the biosphere*. Nachtwey DS, Caldwell MM and Biggs RH (Eds.), CIAP Monograph 5(1), 5/109-166.

Zheng T, Taylor Mayne S, Boyle P, Holford TR, Lui WL and Flannery J (1992). Epidemiology of non-Hodgkin's lymphoma in Connecticut. *Cancer* 70, 840-849

Ziska LH, Teramura AH, Sullivan JH and McCoy A (1993) Influence of ultraviolet-b (UV-B) radiation on photosynthetic and growth characteristics of field grown cassava (*Manihot esculentum* Crantz) *Plant, Cell & Environment* 16, 73-79.

Glossary

UVA,UVB,UVC,UVR:	Ultravviolet (UV) radiation is the region of the electromagnetic spectrum adjacent to that of visible light, lying beyond the shortest visible wavelenghts which are to be found at the violet end of the visible spectrum. UV radiation can be subdivided into 3 wavebands called UVA (wavelength range 315-400nm), UVB (280-315nm) and UVC (200-280nm).
Electromagnetic spectrum:	The range of radiation that includes radio waves, microwaves, infra-red, visible light, ultraviolet, X-rays and gamma rays, which differ from each other only in their wavelengths.
Action Spectrum:	For each biological effect, the relative importance of each wavelength can be expressed in the form of a function known as the action spectrum. Eg. if the action spectrum had a value of 1.0 at 300nm and 0.5 at 320nm, then twice as much energy would be required at 320nm as at 300nm to have the same effect.
Chromophore:	A molecule or group of molecules capable of absorbing light.
Erythema :	reddening of the skin/ sunburn
Montreal Protocol:	The Montreal Protocol on Substances that Deplete the Ozone Layer is an International agreement initially negotiated in 1987 with the objective of eventually eliminating emissions of all ozone depleting substances by progressively phasing out their production and consumption.
Troposphere:	Lowest region of the atmosphere where temperature decreases with height
Stratosphere:	The layer above the troposphere, where the majority of the atmospheric ozone resides
Ozone:	O_3
Total/ column ozone:	The total amount of ozone in a vertical column through the atmosphere
CFCs:	Chlorofluorocarbons, these comprise a key group of chemicals responsible for depleting stratospheric ozone.
DNA:	Deoxyribonucleic acid, Basic genetic material.
Free radical:	A reactive atom or molecule with an odd number of electrons.

Annex 2

Membership of UVB Impacts and Measurements Review Group

UV Measurements and Impacts Review Group (UMIRG) Draft terms of reference

To review current knowledge of the climatology of the biologically damaging UVB radiation of the United Kingdom and the environmental impacts of current levels of UVB; to assess the potential increases in UVB as a result of stratospheric ozone depletion over the UK and, building on published work, review the potential impacts of these; and to advise the Department on further research required in UVB measurements and impacts areas in the UK.

The purpose of the Review Group is therefore to provide the Department of the Environment with the best current scientific assessment of UVB measurement and impacts for the UK on which to base policy; and a thorough assessment of the severity of the consequences should current policies fail.

Dr Ann Webb
Department of Physics
UMIST
PO Box 88
Sackville Street
Manchester M60 1QD

Dr Brian Gardiner
British Antarctic Survey
High Cross
Madingley Rd
Cambridge
CB3 0ET

Dr Colin Driscoll
National Radiological Protection Board
Didcot,
Chilton,
Oxon. OX11 0RQ

Dr Nigel Paul
Inst of Environ & Biol Sciences
Div of Biological Sciences
University of Lancaster
Bailrigg
Lancaster LA1 4YQ

Dr Andy McLeod
ITE Monks Wood
Abbots Ripton
Huntingdon
Cambs. PE17 2LS

Prof John Raven, FRS
Dept of Biol Sciences
University of Dundee
Dundee DD1 4HN

Dr David Lowe
Plymouth Marine Laboratory
Citadel Hill
Plymouth PL1 2PB

Dr J Eric Hillerton
Institute of Animal Health
Compton Laboratory
Newbury
Berkshire RG20 7NN

Prof Brian L Diffey
Regional Medical Physics Dept
Dryburn Hospital
Durham DH1 5TW

Dr Antony R Young
Photobiology Department
Inst of Dermatology
St Thomas's Hospital
London SE11 4TH

Dr Colin Arlett
MRC Cell Mutation Unit
University of Sussex
Falmer
Brighton BN1 9QG

Dr John Austin
Meteorological Office, Room 253
London Road
Bracknell
Berks RG12 2SZ

Dr Duncan Gardiner
Building Research Establishment
Garston
Watford WD2 7JR

Dr Graham Bentham
CSERGE, School of Environ Sciences
University of East Anglia
Norwich NR4 7TJ

Dr Penny Bramwell
Department of the Environment
Room B246
Romney House
43 Marsham Street
London. SW1P 3PY

Experts consulted

Full report
Chairperson: A. Webb
Person consulted: Dr Van der Leun

Chapter 2 Measurements and predictions of ozone and uv
Lead authors: A. Webb; J. Austin; C. Driscoll & B. Gardiner
People consulted: J. Pyle; K. Shine; J. Haigh.

Chapter 3 Molecular effects
Lead author: C. Arlett
Persons consulted: R. Tyrell; M. Noval.

Chapter 4 Aquatic aspects
Lead authors: D. Lowe; J. Raven
People consulted: K. Kramer; D. Hader.

Chapter 5 Impacts on Terrestrial Ecosystems
Lead authors: N. Paul & A. McCleod.
People consulted: T. Callaghan; J. Lee; P. Lumsden

Chapter 6 Human Health Impacts
Lead authors: B. Diffey; A. Young
People consulted: M. Noval; R. McKie, R. Cartwright & J. Marshall.

Chapter 7 Impact on Animals
Lead author: J. E. Hillerton
Person consulted: A. Andrews.

Chapter 8 Impacts on Materials
Lead author: D. Gardiner
People consulted: N. Allen; J. White.

Chapter 9 Air Quality
Lead author: D. Derwent; J. Austin
People consulted: S. Penkett; K. Law

Chapter 10 Socioeconomic impacts
Lead author: G. Bentham
People consulted: A. McMichael

Printed in the United Kingdom
for the Stationery Office
Dd302934 C7 11/96